Advances in

Physical Organic Chemistry

Advances in
Physical Organic Chemistry

Edited by

V. GOLD

Department of Chemistry
King's College, University of London

VOLUME 10

1973

Academic Press, London and New York

A subsidiary of Harcourt Brace Jovanovich, Publishers

ACADEMIC PRESS INC. (LONDON) LTD
24–28 Oval Road, London NW1

U.S. Edition published by
ACADEMIC PRESS INC.
111 Fifth Avenue,
New York, New York 10003

Library of Congress Catalog Card Number: 62–22125
ISBN 0–12–033510–7

PRINTED IN GREAT BRITAIN BY
WILLIAM CLOWES & SONS LIMITED
LONDON, COLCHESTER AND BECCLES

CONTRIBUTORS TO VOLUME 10

D. BETHELL, *The Robert Robinson Laboratories, University of Liverpool, Liverpool, England.*

M. R. BRINKMAN, *The Robert Robinson Laboratories, University of Liverpool, Liverpool, England.*

P. W. CABELL-WHITING, *Department of Organic Chemistry, The University, Zernikelaan, Groningen, The Netherlands.*

R. E. CARTER, *Organic Chemistry 2, Lund Institute of Technology, Chemical Center, S-220 07 Lund 7, Sweden.*

M. FLEISCHMANN, *Department of Chemistry, The University, Southampton SO9 5NH, England.*

H. HOGEVEEN, *Department of Organic Chemistry, The University, Zernikelaan, Groningen, The Netherlands.*

L. MELANDER, *Department of Organic Chemistry, University of Göteborg and Chalmers Institute of Technology, S-402 20 Göteborg 5, Sweden.*

D. PLETCHER, *Department of Chemistry, The University, Southampton SO9 5NH, England.*

CONTENTS

CONTRIBUTORS TO VOLUME 10 v

Experiments on the Nature of Steric Isotope Effects

ROBERT E. CARTER and LARS MELANDER

I. Introduction 1
II. Theory 5
  A. Bartell's Perturbation Treatment 5
  B. An Alternative, Somewhat Naïve and Unsatisfactory
    Model 10
III. Survey of Experimental Results 14
  References 26

The Reactivity of Carbonium Ions Towards Carbon Monoxide

H. HOGEVEEN

I. Introduction 29
II. Reactivity and Stabilization of Tertiary Alkyl Cations . 31
III. Reactivity and Stabilization of Secondary Alkyl Cations . 34
  A. Interconversion of Tertiary and Secondary Pentyloxo-
    carbonium Ions 34
  B. Rate Expressions for the Rearrangement-Carbonylation 37
  C. Interconversion of Tertiary and Secondary Butyloxo-
    carbonium Ions 40
  D. Kinetic and Thermodynamic Control in the Reversible
    Carbonylation of the 2-Norbornyl Cation . . . 41
IV. Interception of Unstable Cations by Carbon Monoxide . 43
  A. Primary Alkyl Cations 43
  B. Vinyl Cations 45
V. Reactivity of Stabilized Carbonium Ions 46
VI. Some Remarks on the Mechanism of Carbonylation . . 50
VII. Conclusions 51
  References 51

Chemically Induced Dynamic Nuclear Spin Polarization and its Applications

D. BETHELL and M. R. BRINKMAN

I. Introduction 53
    A. Scope 53
    B. The N.M.R. Experiment 54
    C. Dynamic Nuclear Polarization 55
II. The Radical Pair Theory of CIDNP 56
    A. Basic Concepts 56
    B. The Relative Energies of Ground-State T and S Manifolds of a Radical Pair 61
    C. The Dynamic Behaviour of Radical Pairs . . . 63
    D. Mechanisms of T–S Mixing 65
    E. Quantitative Aspects of T_0–S Mixing 68
III. Qualitative Analysis of CIDNP Spectra 73
IV. CIDNP at Low Magnetic Fields 76
V. Applications of CIDNP 78
    A. Scope and Limitations 78
    B. Reactions of Diacyl Peroxides and Related Compounds . 82
    C. Reactions of Azo-, Diazo-, and Related Compounds . 95
    D. The Photolysis of Aldehydes and Ketones . . . 104
    E. Reactions Involving Organometallic Compounds . . 110
    F. Ylid and Related Molecular Rearrangements . . 115
VI. Chemically Induced Dynamic Electron Spin Polarization (CIDEP) 120
VII. Prospects 121
    References 122

The Photochemistry of Carbonium Ions

P. W. CABELL-WHITING and H. HOGEVEEN

I. Introduction 129
II. Valence Isomerization Reactions 130
    A. Non-benzenoid Aromatics 130
    B. Alkylbenzenium Ions 133
    C. Protonated Cyclohexadienones 137
    D. Heteroaromatics 139
    E. Protonated Cycloheptadienones 142
III. Electron Transfer and Coupling Reactions . . . 145
    A. Cyclopropenyl Cations 145
    B. Triphenylmethyl (Trityl) Cation 145

IV. Conclusions 150
    References 151

Physical Parameters for the Control of Organic Electrode Processes

M. Fleischmann and D. Pletcher

  I. Introduction 155
 II. Electrode Potential 156
     A. The Electron Transfer Process 157
     B. The Adsorption Equilibria 165
     C. The Electrode Surface 171
III. The Solution Environment 172
     A. Basic Requirements 173
     B. The Environment as a Reactant . . . 174
     C. pH Effects 178
     D. Double Layer and Adsorption Effects 184
 IV. Electrode Material 191
  V. Substrate Concentration 198
 VI. Temperature 201
VII. Pressure 204
VIII. Structural Effects 206
     A. The Rates of Simple Electrode Reactions . . . 206
     B. Reactions of Intermediates 210
     C. Reaction Coordinates 211
 IX. Cell Design 213
     References 220

EXPERIMENTS ON THE NATURE OF STERIC ISOTOPE EFFECTS

ROBERT E. CARTER

*Organic Chemistry 2, Lund Institute of Technology, Chemical Center,
S-220 07 Lund 7, Sweden*

AND

LARS MELANDER

*Department of Organic Chemistry, University of Göteborg and Chalmers
Institute of Technology, S-402 20 Göteborg 5, Sweden*

I. Introduction	.	1
II. Theory	.	5
A. Bartell's Perturbation Treatment	.	5
B. An Alternative, Somewhat Naïve and Unsatisfactory Model	.	10
III. Survey of Experimental Results	.	14
References	.	26

I. INTRODUCTION

THE notion that the structure and conformation(s) of molecules are determined by valence forces in conjunction with non-bonded forces of varying importance is widely accepted by organic chemists. Although the fact that strong crowding of matter in space will give rise to repulsive forces has always been understood in a qualitative manner, and has prompted suggestions of "steric effects" as simple explanations of more or less unexpected phenomena, it is only recently that the implications of non-bonded forces for the conformations of simple molecules, not generally considered overcrowded, have been recognized by chemists. In molecular spectroscopy, on the other hand, non-bonded interactions in such molecules have been taken into account for the past forty years. Thus, the vibrational behaviour of tetrahedral molecules has been treated quantitatively in terms of the so-called Urey-Bradley potential function, which is a combination of a valence-force function and a central-force function. The interaction of neighbouring non-bonded atoms is described by the central-force function. (See, for instance, Wilson *et al.*, 1955.) The quantitative theory of steric isotope effects as developed by Bartell (1961a, b) grew out of his experience with such ideas as a molecular spectroscopist.

1

Steric isotope effects may be ascribed to differences in "effective size" of isotopic atoms. The early part of our discussion will be concerned with the problem of the meaning of this concept. The experimental results which are to be explained are differences in the positions of chemical equilibria and in reaction velocities arising from this difference in "size".

The steric environment of the atoms in the vicinity of the reaction centre will change in the course of a chemical reaction, and consequently the potential energy due to non-bonded interactions will in general also change and contribute to the free energy of activation. The effect is mainly on the vibrational energy levels, and since they are usually widely spaced, the contribution is to the enthalpy rather than the entropy. When low vibrational frequencies or internal rotations are involved, however, effects on entropy might of course also be expected. In any case, the rather universal non-bonded effects will affect the rates of essentially all chemical reactions, and not only the rates of reactions that are subject to obvious "steric effects" in the classical sense.

Isotopic molecules will have force fields which are identical to a high degree of accuracy. The vibrational amplitudes, on the other hand, will be mass-dependent, which means that the steric requirements of isotopic molecules will be slightly different. For this reason it is to be expected quite generally that isotopic molecules will respond differently to the change in steric conditions imposed by a chemical reaction, and hence that their reaction rates will differ somewhat.

In all non-bonded interactions which we shall discuss below, at least one of the atoms belongs to the element of hydrogen, and it is always hydrogen linked to carbon which is isotopic and consists of either protium or deuterium. It is frequently stated that protium "requires more space" than deuterium, but it is worth while examining this statement in detail.

For our purposes, it is satisfactory to use the perfect harmonic oscillator in its ground state as a model for the description of the oscillation of the carbon–hydrogen bond distance around its mean value and of the oscillation of the hydrogen atom around its mean position, all in the absence of non-bonded interactions. Anharmonicity effects can thus be neglected. The almost perfect mass-independence of the electronic structure means that one and the same parabola can be used to describe the potential, irrespective of the hydrogen mass. This implies that the *average* length is the same for carbon-protium and carbon–deuterium bonds. Owing to the difference in reduced mass the probability distribution function will be somewhat broader for the carbon–protium than for the carbon–deuterium oscillator. The probability of a particularly *long* carbon–hydrogen distance is therefore larger with protium than with deuterium. This is probably the only argument in favour of the concept of a greater spatial requirement of the carbon–protium oscillator, which is valid independent of the nature of the perturbing non-bonded potential. It must be borne in mind that, owing to the assumed symmetry of the oscillator potential, the carbon–protium oscillator also has an equal, high probability for a particularly *short* interatomic distance and surpasses carbon–deuterium in this respect as well. This could under some circumstances compensate for the high probability of long distances, and, as already mentioned above, the average distance is actually independent of mass. It is obvious that the concept of a difference in spatial requirements is not founded on quite such simple principles as might be thought at a first glance. In the theoretical section it will be shown that the reason why this concept leads to useful predictions arises from the fact that the even derivatives of the perturbing non-bonded potential are positive. Such perturbations have a stronger effect on the carbon–protium than on the carbon–deuterium oscillator owing to the broader distribution function of the former.

Studies of the molar volumes of perdeuteriated organic compounds might be expected to be informative about non-bonded intermolecular forces and their manifestations, and such studies might be considered to obviate the necessity of investigating steric isotope effects in *reacting* systems. The results from non-reacting systems could then be simply applied to the initial and transition states in order to account for a *kinetic* steric isotope effect.

The molar volumes of CH_4 and CD_4 are known to differ in the expected direction in the condensed phases at low temperatures (Clusius and Weigand, 1940; Grigor and Steele, 1968). Such a difference would in principle originate in the movement of the molecule as a whole as well as in its carbon–hydrogen oscillations, but a macroscopically available compound would be expected to offer better possibilities for experimental investigation than poorly known transition states.

Besides the general feeling that no prediction is so safe that experimental tests can be dispensed with, there is one reason in particular that a thorough experimental investigation of kinetic steric isotope effects seems worth while, at least in the opinion of the present authors. In a kinetic study of these effects two different time scales may be said to meet, that of ordinary intramolecular atomic vibration and that of passage through the transition state along the reaction coordinate. Although the experimental outcome has been entirely in agreement with expectations based on the ordinary transition-state model, somewhat different assumptions would lead to a different prediction, as explained in Section II, B.

In order for an experimental test of the kinetic behaviour to be as informative as possible, the system investigated should fulfil various specific requirements. From the experimental point of view, the reaction should cause a minimum of change in the reaction medium and be without side-reactions as far as possible, in order for accurate and well-defined rate measurements to be feasible. For the same reason an accurate physical method which can be applied without disturbing the reacting system is to be preferred. From the theoretical point of view, it is desirable that the steric effects play as important a role in the reaction as possible, because only then is a sizeable effect to be expected. Finally, a transition state of well-known conformation is a necessary prerequisite for the quantitative application of the theory.

The configurational inversion of biphenyls with various kinds of hindrance has received much attention in recent years. Mislow *et al.* (1964) were first to study steric isotope effects in this context (see Section III). The type of reaction complies with most of the above prerequisites, perhaps to such a degree that it can hardly be considered a true chemical reaction in the usual meaning of the term. On the other hand, it undoubtedly allows the study of steric hindrance as a rate-determining energy threshold. The requirement of a transition state of known conformation brought to mind the fact that the transition state for the configurational inversion of 2,2'-dibromo-4,4'-dicarboxybiphenyl had been treated in great detail first by Westheimer and Mayer (1946) and by Westheimer (1947). Furthermore, Howlett (1960) had subsequently used a different H $\cdots$ Br potential function to calculate the activation energy for inversion of 2,2'-dibromobiphenyl, in a general study of a number of *ortho*-halogenated biphenyls. Thus there were two independent theoretical calculations available on which a comparison between theory and experiment could be based. In 2,2'-dibromo-4,4'-dicarboxybiphenyl, the steric hindrance consists of the repulsion between the two bromine atoms and the hydrogen atoms occupying the other two positions *ortho* to the pivot bond. In order for inversion to take place, a rather strongly distorted (planar) transition state must be passed, in which an appropriate compromise is reached between skeletal deformation and hydrogen–bromine interpenetration. Studies of steric isotope effects in biphenyl inversion had not yet been carried out with the isotopic hydrogen in positions so well defined geometrically as the aromatic ones, and consequently it seemed worth while to undertake the somewhat laborious synthesis of 2,2'-dibromo-4,4'-dicarboxybiphenyl-6,6'-d_2 (1) (Melander and Carter, 1964). Although it was impossible to separate the enthalpy and entropy effects, the observed isotope effect on the free energy of activation was in good agreement with the prediction of Bartell's (1961a) theory when applied to Howlett's (1960) transition-state model if a negligible isotope effect on the entropy was assumed.

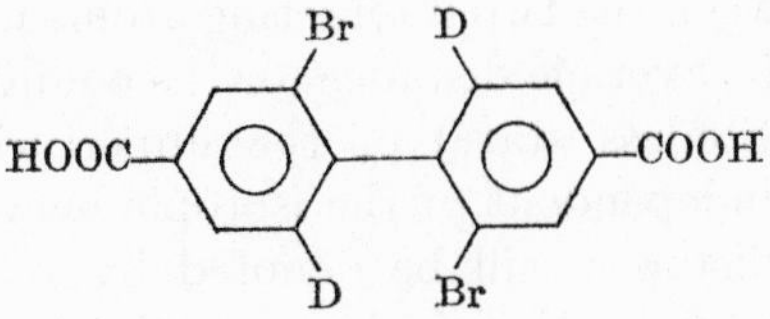

(1)

In the following, a detailed exposition of Bartell's (1961a) theory of steric isotope effects will be given (Section II, A), and an alternative model will be developed, based on somewhat different assumptions about the timing in the transition state, which leads to predictions at variance with the experimental results (Section II, B). In both of these sub-sections, special reference will be made to the work of Melander and Carter (1964). Finally, a selective non-comprehensive review of other experimental work in this field will be presented (Section III).

II. Theory

A. Bartell's Perturbation Treatment

Bartell (1961a) has demonstrated the use of first-order perturbation theory in finding the isotopic difference in non-bonded interaction energy for hydrogen/deuterium (bonded to carbon) in contact with another atom. Using this method to compute the isotopic difference in a transition state implies the assumption that the transition-state lifetime is sufficiently long to allow time-averaging. The fact that the method is capable of making predictions which are quite compatible with experimental results, especially if the inherent uncertainties are taken into account, may be considered evidence in favour of our ordinary way of dealing with transition states.

According to Bartell (1961a), the relative motion of the interacting non-bonded atoms is described by means of a harmonic oscillator when the two atoms are bonded to the same atom, and by means of two super-imposed harmonic oscillators when the atoms are linked to each other via more than one intervening atom. It is the second case which is of interest in connection with the biphenyl inversion transition state. The non-bonded interaction will of course introduce anharmonicity, but since a first-order perturbation calculation of the energy only implies an averaging of the perturbation over the unperturbed state, the harmonic approximation will suffice. Since each oscillator may be assumed to be in its ground state, the natural anharmonicity that exists in the absence of the non-bonded interaction may also be neglected.

The distance between the two interacting atoms will be denoted by r, and r_g will denote the average distance (or the equilibrium distance in a hypothetical vibrationless state) of the unperturbed system. (The distance r_g will be independent of the isotopic mass.) A displacement from the average distance will be denoted by $x = r - r_g$. The two-oscillator model subdivides the displacement into two components: x_m corresponds to the motion of the isotopic hydrogen relative to the carbon atom to which it is attached, and x_s corresponds to the motion of the remainder of the molecular framework joining the two non-bonded atoms. The motion of the isotopic hydrogen relative to the carbon will be sensitive to the isotopic mass, whereas the framework motion will not. The two kinds of displacement are assumed to be governed by the probability distribution functions

$$P_m(x_m) = (2\pi\overline{x_m^2})^{-1/2}\exp\left(-x_m^2/2\overline{x_m^2}\right) \tag{1}$$

$$P_s(x_s) = (2\pi\overline{x_s^2})^{-1/2}\exp\left(-x_s^2/2\overline{x_s^2}\right) \tag{2}$$

In these expressions $\overline{x_m^2}$ and $\overline{x_s^2}$ are the mean squares of the displacements x_m and x_s, and $\overline{x_m^2}$ is thus mass-sensitive while $\overline{x_s^2}$ is not. The mean square $\overline{x^2}$ of the total displacement will be the sum of these two mean squares as easily seen in the following way:

$$\overline{x^2} = \overline{(x_m + x_s)^2} = \int\limits_{-\infty}^{+\infty}\int\limits_{-\infty}^{+\infty} (x_m + x_s)^2\, P_m(x_m)\, P_s(x_s)\, dx_m\, dx_s$$

$$= \int\limits_{-\infty}^{+\infty} x_m^2\, P_m(x_m)\, dx_m + \int\limits_{-\infty}^{+\infty} x_s^2\, P_s(x_s)\, dx_s = \overline{x_m^2} + \overline{x_s^2}. \tag{3}$$

(Since the probability distribution functions (1) and (2) are even functions of x_m and x_s, respectively, the cross-term integral containing x_m and x_s to the first power will vanish.)

(In Bartell's original paper (1961a) the symbols l_m, l_s and l_t are used in the sense of $(\overline{x_m^2})^{1/2}$, $(\overline{x_s^2})^{1/2}$ and $(\overline{x^2})^{1/2}$.)

The non-bonded interaction is now introduced as a perturbation $V(r)$ on this double oscillator, and the perturbation energy becomes

$$\overline{V}(r_g) = \int\limits_{-\infty}^{+\infty}\int\limits_{-\infty}^{+\infty} V(r)\, P_m(x_m)\, P_s(x_s)\, dx_m\, dx_s \tag{4}$$

It is convenient to expand $V(r)$ in a Taylor series around r_g:

$$V(r) = V(r_g) + x V'(r_g) + \tfrac{1}{2}x^2 V''(r_g) + \tfrac{1}{6}x^3 V^{iii}(r_g) + \tfrac{1}{24}x^4 V^{iv}(r_g) + \ldots$$
(5)

(In Bartell's treatment many simultaneous pair-wise interactions are accounted for by $V(r) = \sum V_{ij}(r_{ij})$, which is correspondingly expanded in several independent variables. This is not necessary in the simple case treated here.)

On introducing (5) into (4), use is made of the fact that odd powers of x_m and x_s will always lead to zero integrals when multiplied by the even functions P_m and P_s. The different powers of x may thus be simply written as follows:

$x = x_m + x_s = $ sum of odd power of x_m and x_s

$x^2 = (x_m + x_s)^2 = x_m^2 + x_s^2 + $ term of odd power in each of x_m and x_s

$x^3 = (x_m + x_s)^3 = $ sum of terms of odd power in either x_m or x_s

$x^4 = (x_m + x_s)^4 = x_m^4 + x_s^4 + 6x_m^2 x_s^2 + $ terms of odd power in each of x_m and x_s

and the non-vanishing part of the perturbation energy becomes

$$\bar{V}(r_g) = V(r_g) + \tfrac{1}{2}V''(r_g)\left[\int_{-\infty}^{+\infty} x_m^2 P_m(x_m)\,dx_m + \int_{-\infty}^{+\infty} x_s^2 P_s(x_s)\,dx_s\right]$$

$$+ \tfrac{1}{24}V^{iv}(r_g)\left[\int_{-\infty}^{+\infty} x_m^4 P_m(x_m)\,dx_m + \int_{-\infty}^{+\infty} x_s^4 P_s(x_s)\,dx_s\right.$$

$$\left. + 6\int_{-\infty}^{+\infty} x_m^2 P_m(x_m)\,dx_m \cdot \int_{-\infty}^{+\infty} x_s^2 P_s(x_s)\,dx_s\right] + \ldots$$

$$= V(r_g) + \tfrac{1}{2}V''(r_g)[\overline{x_m^2} + \overline{x_s^2}]$$

$$+ \tfrac{1}{24}V^{iv}(r_g)[3(\overline{x_m^2})^2 + 3(\overline{x_s^2})^2 + 6\overline{x_m^2}\,\overline{x_s^2}] + \ldots$$

$$= V(r_g) + \tfrac{1}{2}\overline{x^2}V''(r_g) + \tfrac{1}{8}(\overline{x^2})^2 V^{iv}(r_g) + \ldots$$
(6)

where relation (3) was used at the end of the calculation.

The conclusion to be drawn from equation (6) is that the perturbation energy is equal to the value of the perturbing potential at the equilibrium separation plus terms which are proportional to the even derivatives of $V(r)$ at the equilibrium separation, and also proportional to increasing powers of the mean square of the total deviation from this separation. It is via this mean square that the isotopic mass will affect the perturbation energy.

When two similar systems, one containing protium and the other deuterium attached to a carbon atom, are subject to the same perturbation $V(r)$ by the approach of a non-bonded atom, we obtain by means of expression (6) the following difference in perturbation energy:

$$
\begin{aligned}
(\overline{V}(r_g))_H - (\overline{V}(r_g))_D &= \tfrac{1}{2} V''(r_g)[\overline{x_H^2} - \overline{x_D^2}] + \tfrac{1}{8} V^{iv}(r_g)[(\overline{x_H^2})^2 - (\overline{x_D^2})^2] + \cdots \\
&= (\overline{x_H^2} - \overline{x_D^2})[\tfrac{1}{2} V''(r_g) + \tfrac{1}{8} V^{iv}(r_g)(\overline{x_H^2} + \overline{x_D^2})] + \cdots \\
&\approx \tfrac{1}{2}(\overline{x_{mH}^2} - \overline{x_{mD}^2})[V''(r_g) + \tfrac{1}{2}\overline{x_H^2} V^{iv}(r_g)]
\end{aligned}
\tag{7}
$$

Relation (3), together with the mass-independence of $\overline{x_s^2}$, gave the first parenthesis in the final expression, and the coefficient of the fourth derivative was obtained from the approximate equality $\overline{x_H^2} + \overline{x_D^2} \approx 2\overline{x_H^2}$. Finally, terms containing higher derivatives than the fourth were neglected.

The isotopic difference obviously depends exclusively on the even derivatives of the perturbing potential, which implies that a linear potential, corresponding to a constant force, gives rise to no isotope effect. This may be easily understood in the following way: if two functions of a variable x, one parabolic ($f_1(x)$) and the other linear ($f_2(x)$), i.e.,

$$
f_1(x) = ax^2 \quad \text{and} \quad f_2(x) = bx + c
$$

are added, the sum

$$
f_1(x) + f_2(x) = ax^2 + bx + c = a\left(x + \frac{b}{2a}\right)^2 - \frac{b^2}{4a} + c
$$

will still represent a parabola of the same shape as $f_1(x)$ but with its vertex moved from the point $x = 0$, $y = 0$ to the point $x = -b/2a$, $y = -(b^2/4a) + c$. Thus the linear potential will only push the parabolic potential well along the x axis and also change the level of its minimum, but its shape and hence the frequency and the positions of the energy levels relative to the classical minimum will not change for any oscillator described by the potential. This principle is of importance in applying Bartell's theory to the biphenyl transition state.

Hitherto it has been assumed that r_g corresponds to the classical equilibrium (or quantum-mechanical average) distance between the non-bonded atoms in the absence of interaction. It is inherent in the proper application of first-order perturbation theory that the perturbation is assumed to be small. In the case of the hindered biphenyls, however, it is known from the calculations cited in the introduction that the transition state is distorted to a considerable extent. The hydrogen atom does not occupy the same position relative to the bromine atom that it would if all the other atoms were still fixed at their positions in the distorted molecule but the hydrogen–bromine repulsion removed. Thanks to the conformational computations we know the classical equilibrium position in the presence of the non-bonded interaction. Hence it seems better to use this result as a starting point and use the values of the second and fourth derivatives of the repulsion potential at the new equilibrium position. This amounts to considering only the second-degree and higher terms of equation (5) as the real perturbation, and r_g is now the new equilibrium distance. The linear part of $V(r)$ with r_g equal to the equilibrium distance in the complete absence of non-bonded interaction can be said to have been already used in the computations of the Westheimer-Mayer type for finding the new equilibrium position. Since neither $V(r_g)$ nor $V'(r_g)$ appears in the final expression (7), the same expression would still result from the last-mentioned type of perturbation calculation, but the difference in the meaning of r_g should be borne in mind.

The isotopic difference between the mean squares of the displacements in equation (7) can be calculated if the carbon–hydrogen oscillator is treated as a diatomic molecule. It is easily shown that for constant potential the mean square of the displacement from the equilibrium position of the harmonic oscillator will be inversely proportional to the square root of the reduced mass, μ, and hence

$$\overline{x^2_{mD}}/\overline{x^2_{mH}} = (\mu_{C-H}/\mu_{C-D})^{1/2} = [(\tfrac{1}{12}+\tfrac{1}{2})/(\tfrac{1}{12}+\tfrac{1}{1})]^{1/2} = (7/13)^{1/2} = 0.734 \tag{8}$$

Relation (8) is used to eliminate $\overline{x^2_{mD}}$ from equation (7), and the result is

$$(\overline{V}(r_g))_H - (\overline{V}(r_g))_D = 0.13_3 \overline{x^2_{mH}}[V''(r_g) + \tfrac{1}{2}\overline{x^2_H}V^{iv}(r_g)] \tag{9}$$

When equation (9) is applied to the transition state of the biphenyl system, it gives directly the isotopic difference in the activation enthalpy per interacting pair of atoms, provided we make the reasonable assumption that initial-state steric effects are independent of isotopic substitution in the 6- and 6'-positions. Since there are two pairs of interacting atoms in the coplanar transition state, the final expression is

$$\Delta H^{\ddagger}_H - \Delta H^{\ddagger}_D = \Delta\Delta H^{\ddagger} = 0.27 \overline{x^2_{mH}}[V''(r_g) + \tfrac{1}{2}\overline{x^2_H}V^{iv}(r_g)] \tag{10}$$

The application of equations (9) and (10) to experimental data will be discussed below in terms of specific forms of the potential function $V(r)$.

The even derivatives of the interaction potential are positive, and thus the prediction in the case of biphenyl inversion is invariably that the enthalpy of activation will be greater for the protium than for the deuterium compound. Since no appreciable effect on the entropy of activation is likely, as will be shown below, the light compound will react at a lower rate than the heavy one.

Using the terminology of Wolfsberg and Stern (1964), the isotopic rate ratio may be expressed as equation (11):

$$k_D/k_H = (MMI) \times (EXC) \times (ZPE) \tag{11}$$

MMI represents the mass and moment-of-inertia term that arises from the translational and rotational partition functions; EXC, which may be approximated to unity at "low" temperatures, arises from excitation of vibrations, and finally ZPE is the vibrational zero-point-energy term. The relation between these terms and the isotopic enthalpy and entropy differences may be written

$$\ln(k_D/k_H) = \ln(ZPE) + \ln(MMI) = (\Delta\Delta H^{\ddagger}/RT) - (\Delta\Delta S^{\ddagger}/R) \tag{12}$$

The ZPE term represents the enthalpy contribution to the isotope effect, and

$$-R\ln(MMI) = \Delta\Delta S^{\ddagger} \tag{13}$$

In terms of principal moments of inertia A, B and C, and the molecular mass M, the entropy term is then given by equation (14) (cf. also Leffek and Matheson, 1971).

$$\Delta\Delta S^{\ddagger} = -\frac{R}{2}\ln\left[\frac{A_H B_H C_H \cdot A_D^{\ddagger} B_D^{\ddagger} C_D^{\ddagger}}{A_D B_D C_D \cdot A_H^{\ddagger} B_H^{\ddagger} C_H^{\ddagger}}\right] - \frac{3R}{2}\ln\left[\frac{M_H M_D^{\ddagger}}{M_D M_H^{\ddagger}}\right] \tag{14}$$

The second term on the right-hand side of this equation vanishes in the case of the configurational inversion of an optically active biphenyl, since the molecular mass does not change between initial and transition states. As for the first term, it seems quite plausible that the introduction of two deuterium atoms in aromatic positions on the biphenyl skeleton will not significantly change the principal moments of inertia. Cancellation will also take place to a large extent between the initial and the transition state. The present model neglects the role of the solvent, but when the isotopic substitution is limited to ordinary aromatic hydrogen, no appreciable isotopic difference in the interaction with the solvent is to be expected. Consequently, the $\Delta\Delta S^{\ddagger}$ term may be safely approximated by zero for the biphenyl cases under consideration here.

B. An Alternative, Somewhat Naïve and Unsatisfactory Model

As already indicated above, kinetic studies of steric isotope effects afford a possibility of testing our present ideas about the transition state. Bartell's perturbation treatment (Section II, A) involves a time-averaging of the repulsion energy with respect to the ordinary vibrational motion within the transition state. The passage through the transition state along the reaction coordinate is thus assumed to be slow compared to the periodic modes orthogonal to that coordinate. Such a picture seems to be in agreement with the commonly accepted view that there exists an approximate equilibrium between the transition state and the reactants. However, the good agreement in general between the predictions from Bartell's method of calculation and the experimental results would have little value as a support for such ideas if different assumptions led to similar predictions. In order to show that the prediction furnished by Bartell's treatment of the transition state in biphenyl inversion is not compatible with every kind of timing within this transition state, an alternative (although admittedly somewhat factitious) model will also be discussed.

As explained in the Introduction, the carbon–protium oscillator is superior to the carbon–deuterium one not only in attaining particularly long bond distances but also in attaining particularly short bond distances. Thus, for instance, the average distance will be the same in both kinds of oscillators. One may therefore ask how the average reaction rates would compare if the passage through the transition state were fast relative to the ordinary vibrational motions. In such a case the protium system might derive advantage from the better capability of its hydrogen atom to get out of the way of the repelling atom.

Let us first consider a single pair of non-bonded atoms which make repeated attempts to pass each other and let us assume that the rate of a single passage is large compared to the rate of the carbon–hydrogen oscillation. The instantaneous carbon–hydrogen distance and the framework distortion determine together the non-bonded distance at which the passage has to take place. The attempted passage will be successful if enough energy is available to overcome the threshold thus dictated. The mass-dependent carbon–hydrogen oscillator will give rise to isotopic differences, and we shall first study its influence on the number of successful passages, the framework distortion being held constant in all passages under consideration.

It will be assumed for the moment that the non-bonded atoms will pass each other at the distance r_g (equal to that found in a Westheimer-Mayer calculation) if the carbon–hydrogen oscillator happens to be in its average position and otherwise at the distance $r = r_g + x_m$, where x_m is a mass-sensitive displacement governed by the probability distribution function (1). The potential-energy threshold felt is assumed to have the value $E(0)$ when $x_m = 0$ and otherwise to be a function $E(x_m)$ which depends on the variation of the non-bonded potential V with x_m

$$E(x_m) = E(0) + V(x_m) - V(0) \tag{15}$$

The fractional probability that the carbon–hydrogen oscillator is within the interval between x_m and $x_m + dx_m$ is $P_m(x_m)dx_m$, and if Z is the total number of attempted passages in unit time, $ZP_m(x_m)dx_m$ will fall in this interval. The fraction of these which will be successful may be assumed to be $\exp[-E(x_m)/kT]$ in the usual way, k being Boltzmann's constant and T the absolute temperature. Hence the total number of successful passages in unit time irrespective of x_m is

$$n = Z \int_{-\infty}^{+\infty} P_m(x_m) \exp[-E(x_m)/kT] \, dx_m$$

$$= Z(2\pi\overline{x_m^2})^{-1/2} \int_{-\infty}^{+\infty} \exp\{-[x_m^2/2\overline{x_m^2}] - [E(x_m)/kT]\} \, dx_m$$

$$= Z(2\pi\overline{x_m^2})^{-1/2} \exp\{-[E(0) - V(0)]/kT\} \int_{-\infty}^{+\infty} \exp\{-[x_m^2/2\overline{x_m^2}] - [V(x_m)/kT]\} \, dx_m$$

where $\overline{x_m^2}$ is the mean square of the oscillator displacement as before. The number Z will be very nearly the same for an ordinary biphenyl and one which is deuteriated in the 2- and 2'-positions. Hence the isotopic ratio for the number of successful passages becomes

$$\frac{n_H}{n_D} = \frac{(\overline{x_{mH}^2})^{-1/2} \int_{-\infty}^{+\infty} \exp\{-[x_m^2/2\overline{x_{mH}^2}] - [V(x_m)/kT]\} \, dx_m}{(\overline{x_{mD}^2})^{-1/2} \int_{-\infty}^{+\infty} \exp\{-[x_m^2/2\overline{x_{mD}^2}] - [V(x_m)/kT]\} \, dx_m} = \frac{\int_{-\infty}^{+\infty} f_H(x_m) \, dx_m}{\int_{-\infty}^{+\infty} f_D(x_m) \, dx_m}$$

(16)

The definite integrals in (16) are not in general expressible in closed form. From separate plots of the integrands of the second-member numerator and denominator, multiplied by $(\overline{x_{mH}^2})^{-1/2}$ and $(\overline{x_{mD}^2})^{-1/2}$, respectively, it is obvious that n_H/n_D exceeds unity by a considerable amount, i.e., the light compound is expected to have a larger number of successful passages per unit time than the heavy one (cf. Fig. 1). (The assumptions underlying Fig. 1 are that $(\overline{x_{mH}^2})^{1/2} = 8 \times 10^{-10}$ cm, corresponding to a wave number of 2833 cm^{-1} for the diatomic carbon–protium oscillator, $\overline{x_{mD}^2}$ has a value obtained from the foregoing via equation (8), $T = 273°K$ and V is the non-bonded potential function used by Howlett (1960), $V(x_m) = 1·44 \times 10^{-10} \exp[-(x_m + r_g)/0·4379]$ erg molecule^{-1}, x_m and r_g being measured in Å units, with $r_g = 2·46$ Å, in agreement with Howlett's computation for 2,2'-dibromobiphenyl.)

The two curves of Fig. 1 intersect when

$$(\overline{x_{mH}^2})^{-1/2} \exp\{-[x_m^2/2\overline{x_{mH}^2}] - [V(x_m)/\mathbf{k}T]\} = (\overline{x_{mD}^2})^{-1/2}$$
$$\exp\{-[x_m^2/2\overline{x_{mD}^2}] - [V(x_m)/\mathbf{k}T]\}$$

or

$$x_m^2[(1/\overline{x_{mD}^2}) - (1/\overline{x_{mH}^2})] = \ln(\overline{x_{mH}^2}/\overline{x_{mD}^2}) \approx (\overline{x_{mH}^2}/\overline{x_{mD}^2}) - 1$$
$$x_m \approx \pm(\overline{x_{mH}^2})^{1/2}$$

and outside the region between the two intersections $f_H(x_m)$ runs above $f_D(x_m)$. Owing to the decrease of $V(x_m)$ with increasing x_m the region $x_m > (\overline{x_{mH}^2})^{1/2}$ gives the predominant contributions to the number of

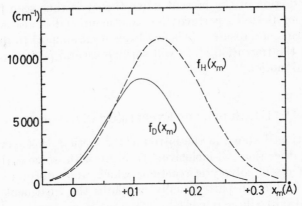

FIG. 1. Diagram for the estimation of the relative magnitudes of the numerator and denominator of the ratio in equation (16).

successful passages, and the protium compound benefits by the some-what larger deviations of protium from its average position. In this case it is obviously the larger probability for the carbon–protium oscillator to attain a particularly short carbon–hydrogen distance that enables the protium atom to get out of the way of the bromine atom more efficiently than the deuterium atom. (In the above reasoning it has been more or less tacitly assumed that all three atoms, carbon, hydrogen and bromine, fall on the same straight line, i.e., that x_m reflects exclusively the carbon–hydrogen stretching mode. The conclusions also hold for the more general and real case in which carbon–hydrogen bending is also partially involved.)

It is a rather arbitrary assumption that the average position of the carbon–hydrogen oscillator should correspond exactly to the distance

r_g between the hydrogen and bromine atoms resulting from a West-heimer-Mayer calculation on a transition state sufficiently long-lived for the total potential energy to attain a minimum. Instead, an averaging with respect to all possible framework distortions as well should be carried out. Moreover, an inversion of the biphenyl would correspond to an attempted one which is successful with respect to both non-bonded pairs of atoms. It is felt, however, that the outcome of such calculations would be qualitatively similar to the present result, because for all degrees of distortion the light compound may be anticipated to have the greatest number of successful passages in unit time at a single point of interaction, and the claim for simultaneous success in two points of interaction would not jeopardize the prediction of a higher rate of inversion for the light compound.

The prediction of the present naïve theory is thus seen to contrast with that of Bartell's perturbation treatment and is also at variance with experimental results. This fact lends some support to the ordinary view that the transition state will survive several periods of ordinary vibrational motion.

III. Survey of Experimental Results

As mentioned above in the introductory section, this survey will be selective rather than all-inclusive. To begin with, some early work on isotope effects in solvolytic reactions which was interpreted in steric terms, and which in part actually pre-dates the treatment of Bartell (1960, 1961a, b) will be reviewed.

Solvolytic reactions do not in general fulfil the most important of the requirements put forward in Section I (p. 4), since (i) they usually involve changes in the medium and significant changes in charge distribution between initial and transition states, (ii) the dominant effect along the reaction coordinate is not necessarily steric in origin, and (iii) the transition state geometry is very seldom unequivocally defined. These facts make it impossible to use the data for even semiquantitative comparisons with the theory, but several papers describing work on secondary kinetic isotope effects in solvolytic reactions will be discussed here because of their historical or general interest.

Leffek, Llewellyn and Robertson (1960a, b) made careful conducto-metric determinations of deuterium kinetic isotope effects on the solvolysis rates (in water) of some ethyl, isopropyl and n-propyl sulphonates and halides. In the case of the n-propyl compounds,

$$CH_3CH_2CH_2X + H_2O \rightarrow CH_3CH_2CH_2OH + HX$$

α-, β- and γ-isotope effects were measured, and it was found that deuterium in the α- or β-position generally slowed down the rate of hydrolysis while deuterium in the γ-position led to an increased rate. Data for the methanesulphonate and the iodide are summarized in Table 1, along with some results by the same authors (Leffek *et al.*, 1960c) on β-isotope effects in ethyl, isopropyl and t-butyl compounds to be discussed later.

TABLE 1

Isotopic Rate Ratios for Various Compounds Hydrolyzing in Water[a]

Compound	Temp., °C	$(k_D/k_H)_\alpha$	$(k_D/k_H)_\beta$	$(k_D/k_H)_\gamma$
n-Propyl methanesulphonate	60·004	0·96$_5$	0·93$_1$	1·06$_1$
n-Propyl iodide	90·003	0·99$_4$	0·93$_9$	1·08$_1$
Ethyl methanesulphonate	60·00$_0$		0·97$_4$	
Ethyl iodide	80·00$_0$		0·96$_5$	
Isopropyl methanesulphonate	30·00$_1$		0·64$_7$	
Isopropyl iodide	60·00$_3$		0·76$_2$	
t-Butyl chloride	2·07$_0$		0·38$_9$	

[a] From Leffek, Llewellyn and Robertson (1960a, b, c); the positions indicated by the subscripts α, β and γ were fully deuteriated.

The α-effect is suggested—independent of the work of Bartell (1960, 1961a,b)—to be "governed largely by steric effects" (Leffek *et al.*, 1960a). That is, van der Waals forces (non-bonded interactions) between the α-hydrogens and the substituent X are reduced as the H(α)-X distance increases on going from the initial to the transition state. The $sp^3 \rightarrow sp^2$ hybridization change on the α-carbon atom between initial and transition states is of course also discussed as a possible source of the observed α-isotope effect (see Streitwieser *et al.*, 1958), but in view of the uncertainties involved in the choice of appropriate force constants for the transition-state vibrations, the evidence that a hybridization change necessarily leads to the change in zero-point energy required to account for the observed effects is regarded as inconclusive.

The effect of γ-deuterium on the hydrolysis of the n-propyl compounds $((k_D/k_H) > 1)$ is interpreted as being entirely due to a reduction in steric interactions involving the terminal methyl group and the leaving group in the transition state of the deuterium compound compared to that of the protium compound (Leffek *et al.*, 1960b). In the light of our present knowledge of secondary deuterium isotope effects, and on the basis of reasonable models for the preferred conformation of the n-propyl chain (see Leffek *et al.*, 1960b), this interpretation appears quite plausible.

The β-deuterium effect was studied and discussed in detail in a later paper (Leffek, Llewellyn and Robertson, 1960c) on the basis of results on the hydrolysis rates of ethyl, isopropyl and t-butyl compounds. (Some examples are included in Table 1.) It is concluded that the β-effect may be rationalized in terms of an interaction between the developing p-orbital on the α-carbon atom and the β-carbon–hydrogen orbital ("hyperconjugation" as described by Kreevoy and Eyring, 1957), and/or in terms of non-bonded interactions involving β-hydrogen atoms. The experiments do not allow a choice to be made between these alternatives.

The dependence of the β-deuterium effect on the spatial orientation of the isotopic bond with respect to the developing p-orbital on the α-carbon atom was elegantly demonstrated by Shiner and Humphrey (1963). This work will not be discussed in detail here; suffice it to say the suggestion is made that β-deuterium effects are better correlated by the postulate of hyperconjugation and its angular dependence than by "the simple steric model" (Shiner and Humphrey, 1963).

Solvolytic experiments specifically designed to test Bartell's theory were carried out by Karabatsos et al. (1967), who were primarily interested in an assessment of the relative contributions of hyperconjugation and non-bonded interactions to secondary kinetic isotope effects. Model calculations of the (steric) isotope effect in the reaction $2 \rightarrow 3$ were performed, as well as that in the solvolyses of acetyl chloride

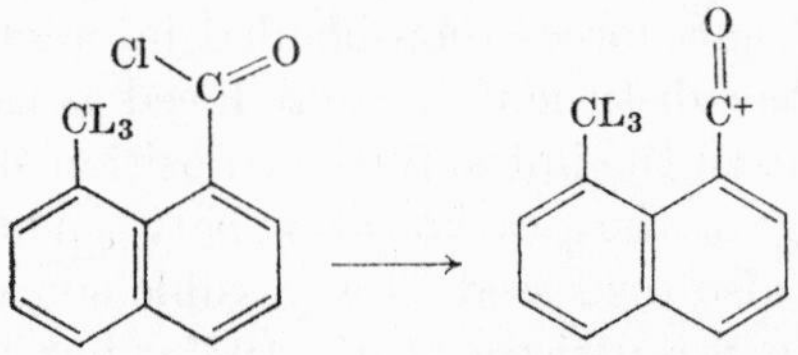

(2)   (3)

L = H, D

and t-butyl chloride, in which hyperconjugation is possible. The calculations were based on the equations of Bartell (1961a) and employed available non-bonded potential functions (Bartell, 1961a; Scott and Scheraga, 1965). The isotopic rate ratios thus obtained were found to be further from unity than the experimental value of $k_D/k_H = 0.972 \pm 0.015$ at 25°C, especially using the potential functions of Bartell (1961a). However, a later paper by two of the same authors (Karabatsos and Papaioannou, 1968) reports a study of the temperature dependence of the isotope effect in the solvolysis of 2, and it is demonstrated that on going from $-20°$ to $-34°$C, the ratio k_D/k_H changes from 0.885 ± 0.027

to $0 \cdot 856 \pm 0 \cdot 009$. The $\Delta\Delta G^{\ddagger} = \Delta G_H^{\ddagger} - \Delta G_D^{\ddagger}$ values at the three temperatures investigated ($-20 \cdot 5°$, $-26 \cdot 6°$, and $-34 \cdot 0°$) are reported as -61 ± 11, -59 ± 8 and -74 ± 4 cal mol^{-1}. A plot of the logarithm of the isotopic rate ratio $vs.$ $1/T$ (including the value at 25°C) provided a $\Delta\Delta H^{\ddagger}$ value of -308 ± 102 cal mol^{-1}, which was compared with calculated values in the range -200 ± 100 cal mol^{-1}. The experimental value of $\Delta\Delta H^{\ddagger}$, in conjunction with $\Delta\Delta G^{\ddagger}$, implies that $\Delta\Delta S^{\ddagger}$ is of the order of -1 cal mol^{-1} deg^{-1}. Karabatsos and Papaioannou (1968) do not comment upon the isotopic entropy difference, but if it is indeed as large as -1 cal mol^{-1} deg^{-1} this is certainly a significant deviation from the *steric* isotope effect theory, which involves the tacit assumption $\Delta\Delta S^{\ddagger} = 0$ (see Section II, A). The very small variation tends, however, to make an assessment of the enthalpy and entropy contributions to $\Delta\Delta G^{\ddagger}$ quite uncertain. Nevertheless, this work demonstrates clearly the advisability of determining a given isotope effect at several temperatures whenever practicable.

The calculated isotope effects for the solvolyses of t-butyl chloride-d_9 (previously estimated by Bartell, 1961a) and acetyl chloride-d_3 were much weaker than the experimentally observed ones. The authors conclude (Karabatsos *et al.*, 1967) that in ordinary systems where hyperconjugation (from the β-position) is possible, the effect of non-bonded interactions accounts for only a small part (less than 10%) of the observed isotope effect.

In addition to work on biaryls to be separately discussed later, there are several reports in the literature of experiments involving kinetic processes that unequivocally establish the steric relationships $H > D$

$$\begin{array}{cc} \overset{\displaystyle H}{\underset{\displaystyle OH}{H_3C-\overset{|}{\underset{|}{C}}-CD_3}} & \overset{\displaystyle D}{\underset{\displaystyle OH}{R-\overset{|}{\underset{|}{C}}-H}} \\ (4) & (5) \end{array}$$

and $CH_3 > CD_3$. For example, a very sensitive method (Horeau, 1961, 1962) for the assignment of absolute configuration to alcohols has been applied to optically active $(+)$-(S)-2-propanol-1,1,1-d_3 (4) (Horeau *et al.*, 1965), and to a series of primary alcohols of known absolute configuration whose optical activity is the result of the introduction of one deuterium atom on the α-carbon (5) (Horeau and Nouaille, 1966). The method is a partial asymmetric alcoholysis based on the reaction between the optically active alcohol and an excess of racemic α-phenyl-butyric anhydride in pyridine solution, followed by the measurement of the optical rotation of recovered α-phenylbutyric acid (Horeau, 1961,

1962). This type of procedure is referred to as a kinetic resolution since the enantiomers of the racemic substrate exhibit different rates of reaction with the optically active compound, i.e. the diastereomeric transition states that arise from differences in e.g. non-bonded interactions have different free energies. Horeau and Nouaille (1966) estimate that a rate difference corresponding to $\Delta\Delta G^{\neq}$ of the order of 0·2 cal mol^{-1} at 25°C could in principle be revealed by this method.

It has been empirically established (Horeau, 1961) that an alcohol with three groups of varying size (S = small, M = medium, L = large) arranged as indicated in **6** leads to recovered α-phenylbutyric acid

$$\text{M}\diagdown_{\text{C}}\diagup^{\cdots\text{OH}}_{\text{S}\diagup\searrow\text{L}}$$

(6)

containing an excess of the (+)-(S)-enantiomer. Thus, the sign of rotation of the recovered acid is related to the absolute configuration of the inducing alcohol. If the absolute configuration of the alcohol is known, the relative sizes S, M and L may be deduced.

It was found that the signs of rotation of the recovered α-phenylbutyric acid corresponded to the known absolute configurations of the deuteriated alcohols if and only if the size relationships $CH_3 > CD_3$ and $H > D$ were valid. In the case of (+)-(S)-2-propanol-1,1,1-d_3 (**4**), the optical yield was between 0·4 and 0·5% (Horeau *et al.*, 1965), corresponding to a $\Delta\Delta G^{\neq}$ value of about 23 cal mol^{-1} at 25°C. For the primary alcohols, quite analogous results were obtained (Horeau and Nouaille, 1966).

Brown and McDonald (1966) provided another type of kinetic evidence for these size relationships by determining secondary kinetic isotope effects in reactions of pyridine-4-d, pyridine-d_5, and various methyl-d_3-pyridines with alkyl iodides. For example, the isotopic rate ratio in the reaction between 4-(methyl-d_3)-pyridine and methyl iodide at 25·0°C in nitrobenzene solution was determined to be $k_D/k_H = 1·001$, while that in the corresponding reaction with 2,6-(dimethyl-d_6)-pyridine was 1·095. (Brown and McDonald (1966) estimate an uncertainty of 1% in the k_D/k_H values.) Furthermore, the isotopic rate ratio in the case of the 2-(methyl-d_3)-compound increased from 1·030 to 1·073 as the alkyl group in the alkyl iodide was changed from methyl to isopropyl, i.e. the isotope effect increased with increasing steric requirements of the alkyl iodide.

It is worth noting that analogous results were obtained and similar conclusions reached on the basis of determinations of the isotope effects on the heats of reaction of the methyl-d_3- and dimethyl-d_6-pyridines with boron trifluoride and with diborane (Brown *et al.*, 1966).

Both sets of results may also be discussed in terms of inductive differences between hydrogen and deuterium (see Halevi, 1963). Brown *et al.* (1966) point out that both the inductive and steric explanations qualitatively predict isotope effects in the same direction, but that "an inductive effect would be expected to operate from the 3 and 4 positions nearly as effectively as from the 2 position". Furthermore, there is no observable isotope effect on the heat of reaction of 2,6-(dimethyl-d_6)-pyridine with the relatively small molecule diborane ($\Delta\Delta H = -20 \pm 18$ cal mol^{-1}), but a significant effect is obtained with the larger molecule boron trifluoride ($\Delta\Delta H = 230 \pm 150$ cal mol^{-1}).

Kaplan and Thornton (1967) determined secondary kinetic isotope effects in two S_N2 displacement reactions, one of which involved quaternization of N,N-dimethylaniline and N,N-dimethyl-d_6-aniline with methyl *p*-toluenesulphonate (CH$_3$OTs). The reactions were carried out in

$$C_6H_5N(CL_3)_2 + CH_3OTs \xrightarrow[L=H,D]{k_L} C_6H_5N^+(CL_3)_2CH_3 + {}^-OTs$$

nitrobenzene solution at 51·29°C, and the rates were measured conductometrically; $k_D/k_H = 1·133 \pm 0·008$.

This result could be qualitatively interpreted as a steric isotope effect, since the methyl groups are more crowded in the transition state than in the initial state. A *qualitative* interpretation in terms of inductive differences between protium and deuterium (see Halevi, 1963) also predicts an isotope effect in the observed direction, but Kaplan and Thornton (1967) emphasize the steric explanation and essentially ignore the inductive alternative. The authors make quantitative comparisons between theory and experiment by using the complete theory of isotope effects (Bigeleisen and Wolfsberg, 1958; Wolfsberg and Stern, 1964), with assumed force constants from analogous molecules to calculate the equilibrium constant for the isotope exchange equilibrium, and by

$$C_6H_5N(CD_3)_2 + C_6H_5N^+(CH_3)_3 \rightleftharpoons C_6H_5N(CH_3)_2 + C_6H_5N^+(CD_3)_2CH_3$$

calculating the zero-point energy contribution on the basis of observed infrared vibrational frequencies. Such comparisons are certainly justified in this case in view of the concerted nature of the reaction, which implies that the force-constant changes for the methyl hydrogens on going from reactants to transition state are most likely quite similar to the force-constant changes between reactants and products (Kaplan and Thornton, 1967). The kinetic rate ratio is thus expected to be intermediate between 1·0 and the value of the isotope exchange equilibrium constant. In this way, the difficulties due to the absence of detailed information about the geometry of the transition state are avoided, but nonetheless conclusions of significance for the interpretation of the details of the kinetic process may be reached.

Kaplan and Thornton (1967) used three different sets of vibrational frequencies to estimate the zero-point energies of the reactants and products of the equilibrium, which provided three different isotope exchange equilibrium constants: 1·163, 1·311 and 1·050. The value 1·311 is considered to be most reasonable, whereas the others are rejected as unrealistic for the case in hand. Calculations using the complete theory led to values that varied from 1·086 to 1·774 for different sets of valence-force constants for the compounds involved.

Changes in observed CH (CD) stretching frequencies on going from reactant to product are found to be too small and in the wrong direction to account for the observed kinetic isotope effect, and the effect is suggested to be due to "increased force constants for lower frequency vibrations, such as for bending" (Kaplan and Thornton, 1967). This is consistent with a steric explanation.

A recent paper by Leffek and Matheson (1971) nicely complements this work, as it describes the results of a careful investigation of the temperature dependence of the kinetic isotope effect in the reaction studied by Kaplan and Thornton (1967). It is found that $\Delta\Delta H^{\neq} = 134 \pm 30$ cal mol^{-1} and $\Delta\Delta S^{\neq} = 0.15 \pm 0.09$ cal mol^{-1} deg^{-1}, demonstrating that the isotope effect is primarily due to an enthalpy difference, and providing support for the steric interpretation suggested by Kaplan and Thornton (1967).

CL$_3$ CL$_3$

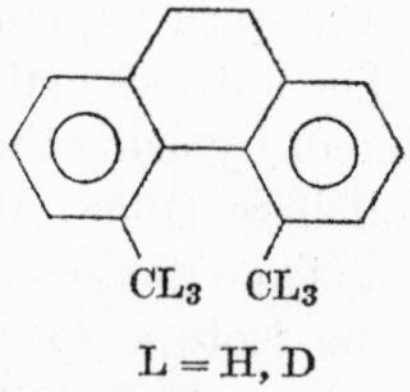

L = H, D

(7)

The molecules most profitably studied in connection with purely *steric* isotope effects have been isotopically substituted biphenyl derivatives. Mislow *et al.* (1964) reported the first more or less clearcut example of this kind in the isotope effect in the configurational inversion of optically active 9,10-dihydro-4,5-dimethylphenanthrene (7), for which an isotopic rate ratio (k_D/k_H) of 1·17 at 295·2°K in benzene solution was determined. The detailed conformation of the transition state is not certain in this case, as it involves the mutual passage of two methyl groups, and thus it is difficult to compare the experimental results with theoretical calculations. Mislow *et al.* (1964) suggest that intermeshing ("the gear effect") is perhaps involved, and this seems eminently reasonable in the light of later (unrelated) work by others. (See, for example, the recent paper by Roussel *et al.*, 1971.) A further complication in the case of **7** is that the introduction of trideuteriomethyl groups will contribute to relief of strain in the initial state, as well as facilitating the passage of the methyl groups in the transition state.

The work of Melander and Carter (1964) on 2,2'-dibromo-4,4'-dicarboxybiphenyl-6,6'-d_2 (**1**) has been referred to above in the introductory and theoretical sections, where it was pointed out that the availability of two detailed theoretical computations of the inversion barrier (Westheimer and Mayer, 1946, Westheimer, 1947; Howlett, 1960) made this system especially attractive for the study of steric isotope effects. Furthermore, in the preferred initial-state conformation the two bromines are probably in van der Waals contact (cf. Hampson and Weissberger, 1936; Bastiansen, 1950), and thus initial-state steric effects are unaffected by deuterium substitution in the 6 and 6' positions. The barrier calculations provided two different theoretical values for the non-bonded $H \cdots Br$ distance in the transition state which, together with the corresponding $H \cdots Br$ potential function, could be inserted in equation (10) to yield values for $\Delta\Delta H^{\ddagger}$. For $(\overline{x_{mH}^2})^{1/2}$ and $(\overline{x_H^2})^{1/2}$ the values 0·09 Å and 0·25 Å were assumed. The value of $\Delta\Delta H^{\ddagger}$ based on Howlett's potential function was found to be 100 cal mol^{-1}, while that based on Westheimer's function was 506 cal mol^{-1}. The former value was encouragingly close to the experimental $\Delta\Delta G^{\ddagger}$ of about 90 cal mol^{-1}. The experimental determinations were carried out in ethanol solution at four temperatures between $-20°$ and $0°C$, yielding isotopic rate ratios in the range 1·19–1·17. It was unfortunately impossible to separate $\Delta\Delta G^{\ddagger}$ into $\Delta\Delta H^{\ddagger}$ and $\Delta\Delta S^{\ddagger}$ components, but the assumption that $\Delta\Delta G^{\ddagger} = \Delta\Delta H^{\ddagger}$, i.e. that there is no isotope effect on the entropy term, seems quite reasonable for the case of a biphenyl inversion (see Section II, A).

An alternative explanation of the results of Melander and Carter (1964) in terms of inductive differences between protium and deuterium was considered unsatisfactory, in view of the insensitivity of the isotopic rate ratio to rather large changes in the properties of the solvent. Although the *rate* of racemization was significantly influenced by a change of solvent (*ca.* a factor of 2 between acetone and 0·5N NaOH), the isotopic rate ratio remained unchanged: $k_D/k_H = 1·18$ in acetone, 1·17 in dimethylformamide and 1·17 in 0·5N aqueous sodium hydroxide solution. Further support for the assumed insignificance of isotopic inductive differences was obtained through the work of Carter and Junggren (1968), who synthesized $(+)$-2,2'-dibromo-4,4'-dicarboxybiphenyl-5,5'-d_2 (8), and found $k_D/k_H = 1 \cdot 02 \pm 0 \cdot 02$ at $-5 \cdot 8°C$ in ethanol solution.

It was of course not possible to introduce the deuterium atoms closer than three bonds away from the "center of reaction", but the attenuation of the inductive effect with distance was not expected to affect the validity of the conclusions, especially considering the work of Streitwieser and Klein (1964), who found that the isotope effect per deuterium in the solvolysis of benzhydryl chloride only decreased from 1·9% for deuterium in the *ortho* positions to 1·5% for deuterium in the *meta* positions.

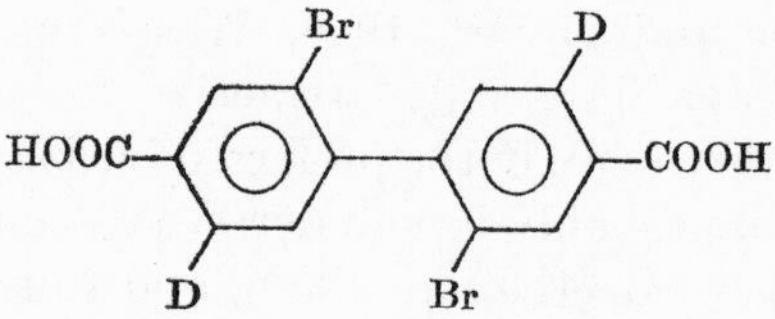

(8)

H · · · H non-bonded interactions are of great importance in organic compounds, and thus it was of interest to attempt to investigate H · · · H non-bonded potential functions via the determination of a *steric* isotope effect in the configurational inversion of an unsubstituted biaryl. In view of the extensive work of Harris and her co-workers in the 1,1'-binaphthyl series (see, for example, Badar *et al.*, 1965; Cooke and Harris, 1963), and since the parent compound is one of the simplest hydrocarbons that may be obtained in enantiomeric forms, the determination of the isotope effect in the inversion of 1,1'-binaphthyl-2,2'-d_2 (9) was

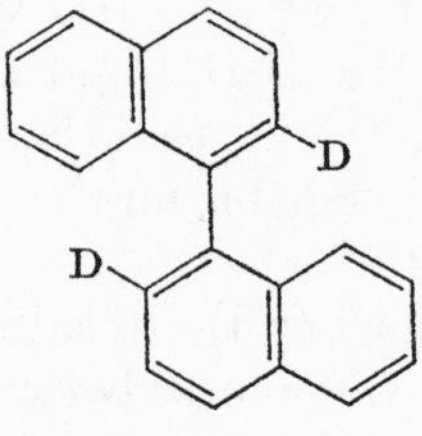

(9)

undertaken (Carter and Dahlgren, 1969). The isotopic rate ratio was obtained at various temperatures between 20° and 65°C in dimethylformamide solution, and an attempt was made to extract $\Delta\Delta H^{\ddagger}$ and $\Delta\Delta S^{\ddagger}$ values from a least-squares treatment of a plot of $\ln{(k_D/k_H)}$ vs. $1/T$. The isotopic rate ratio varied from 1·20 to 1·14, and the values obtained for $\Delta\Delta H^{\ddagger}$ and $\Delta\Delta S^{\ddagger}$ were 270 ± 140 cal mol^{-1} and $0 \cdot 54 \pm$

0·43 cal mol^{-1} deg^{-1}, respectively. The error limits are given as three times the standard deviation obtained from the least-squares calculation, and represent 98% confidence limits. Bearing in mind the theoretical rationalization given at the end of Section II, A for the assumption $\Delta\Delta G^{\ddagger} = \Delta\Delta H^{\ddagger}$, the apparently somewhat arbitrary choice of 100 cal mol^{-1} as a rough value for $\Delta\Delta H^{\ddagger}$ to be used in estimating the non-bonded H $\cdots$ H transition-state distance in the binaphthyl inversion (Carter and Dahlgren, 1969) receives some justification. The transition-state geometry is not well-defined in this case, and thus it was not possible to make a theoretical calculation of the isotope effect as outlined in Section II, A. Harris and co-workers (Badar *et al.*, 1965; Cooke and Harris, 1963) have considered the problem of non-planar transition states in the 1,1′-binaphthyl series, and conclude that considerable bending of the inter-annular bond probably occurs. They suggest that the reaction path for inversion may lead through two discrete transition states, one for each pair of opposing hydrogen atoms. The alternative is a one-stage reaction in which both pairs of atoms pass each other simultaneously. To obtain an estimate of the H $\cdots$ H distance in the transition state(s), Carter and Dahlgren (1969) used the second and fourth derivatives of a variety of H $\cdots$ H non-bonded potential functions from the literature in conjunction with equation (9) for the two-stage case or equation (10) for the one-stage case, and calculated the values of r_g appropriate for a $\Delta\Delta H^{\ddagger}$ value of 100 cal mol^{-1}. The parameters $(\overline{x_{mH}^2})^{1/2}$ and $(\overline{x_H^2})^{1/2}$ were assumed to be 0·09 Å and 0·25 Å, as in the work on 2,2′-dibromo-4,4′-dicarboxy-biphenyl-6,6′-d_2 (Melander and Carter, 1964) referred to above. The values of r_g for 1,1′-binaphthyl derived from this procedure varied from 1·7 Å to 2·2 Å for the two-stage reaction path, and from 1·9 Å to 2·4 Å for the one-stage path. An attempted computer calculation of the geometry of a transition state for the 1,1′-binaphthyl inversion (Carter and Svensson, unpublished), using a well-known minimization routine together with non-bonded potential functions from the work of Scott and Scheraga (1965) and appropriate values for bending and stretching force constants, led to a strongly distorted structure with an unreasonably high energy of distortion and interpenetration (89 kcal mol^{-1}). The energy of the entire system had been minimized with no assumptions made about the symmetry of the deformations involved in the inversion process, in order to allow for an inversion path involving two discrete transition states. Thus, for example, the H $\cdots$ H non-bonded distances for the two pairs of passing atoms were not assumed to be identical. The distances obtained were 1·57 Å and 1·62 Å, which may be compared with values of 2·0 Å (two-stage process) and 2·1 Å (one-stage process) calculated by Carter and Dahlgren (1969) on the basis of $\Delta\Delta H^{\ddagger} = 100$ cal mol^{-1}

and the Scott-Scheraga (1965) H $\cdots$ H non-bonded potential function. Even if both of the computer-calculated non-bonded distances could somehow be "stretched" to 2 Å, the total energy would only decrease to about 64 kcal mol^{-1}, which is still very high in comparison with the observed $\Delta H^{\ddagger}$ of $21\cdot49 \pm 0\cdot19$ kcal mol^{-1} for the protium compound (Carter and Dahlgren, 1969). The computer calculations were not further pursued, as they were judged too difficult and uncertain to be worth while.

In apparent contradiction with Bartell's (1960, 1961a, b) theory, Heitner and Leffek (1966) reported the absence of an isotope effect in

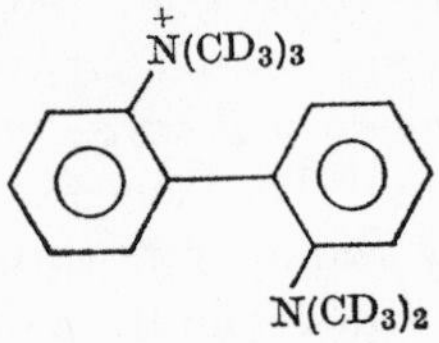

$$\overset{+}{N}(CD_3)_3$$

$$N(CD_3)_2$$

(10)

the racemization of the (+)-2-(N,N,N-*tris*-trideuteriomethyl)-2'-(N,N-*bis*-trideuteriomethyl)diaminobiphenyl cation (10), in spite of the presence of fifteen deuterium atoms in the molecule, and proposed an explanation in terms of compensatory $\Delta\Delta H^{\ddagger}$ and $\Delta\Delta S$ terms. Subsequently, Carter and Dahlgren (1969), on the basis of studies of a molecular model, made the suggestion that the hydrogen atoms of the N-methyl groups may not be involved in the non-bonded interactions that dominate the path to configurational inversion, but that the most important interactions may be between the 6- and 6'-protons and the nitrogen and

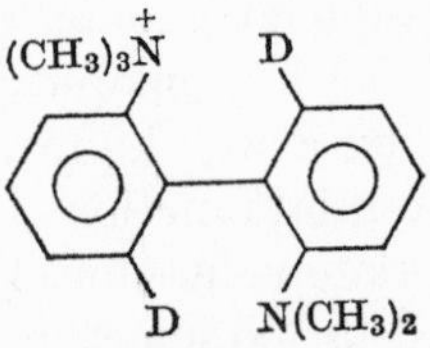

$$(CH_3)_3\overset{+}{N} \quad D$$

$$D \quad N(CH_3)_2$$

(11)

carbon atoms of the 2- and 2'-substituents. The synthesis of the 6,6'-dideuterio compound (11) was undertaken in order to test this suggestion. An isotope effect was indeed observed in this case: the rate ratio k_D/k_H was found to be $1\cdot18 \pm 0\cdot04$ at 98°C and $1\cdot16 \pm 0\cdot05$ at 73°C in aqueous solution (Carter and Dahlgren, 1970). This investigation may be considered to be an application of steric isotope effects as a probe to find the origin of steric hindrance, i.e. to trace the point of non-bonded contact, in an intramolecular collision. Such an application parallels to some extent the generally accepted use of primary isotope effects to trace the bond undergoing rupture in a transition state, or to establish the rate-determining step in a multi-step reaction. In the case of **11**, it may be surmised that as the aryl proton passes the N,N,N-trimethylammonium group, it "intermeshes" among the methyl groups on the nitrogen atom. This represents another type of "gear effect" (*vide supra*; cf. Roussel *et al.*, 1971). In this way, H · · · N interactions (and perhaps H · · · C interactions) become the dominant sources of the potential-energy barrier to inversion.

At first glance it may seem surprising that rather similar k_D/k_H values have been found in the various cases studied. The difference in observation temperature is not too large, and thus the $\Delta\Delta G^{\pm}$ values are also roughly equal. If there is no isotope effect on $\Delta S^{\pm}$, which seems likely, the constancy would apply to $\Delta\Delta H^{\pm}$ as well. The reason for this is probably the fact that our observations have all been made by the same technique, allowing measurements only within a rather narrow rate interval. The systems studied were therefore chosen because they have free energies of activation of a convenient magnitude, and this will tend to equalize the enthalpies of activation. The molecular frameworks have all been rather similar in kind, and hence the degree of distortion in the transition states has probably been about the same. Since the distortion is closely related to the repulsive force between the non-bonded atoms, this force has been of approximately constant magnitude. The ratio between two consecutive derivatives of the usual non-bonded potential functions is generally found to be roughly independent of the kind of interaction, and consequently a similarity between the first derivatives (a similarity of the forces) should be accompanied by a similarity between the second and subsequent derivatives, as well. According to equation (10), this explains the homogeneity of the experimental results.

Howlett (1960) gives a list of non-bonded interaction potentials of the form $V(r) = A\exp(-r/\rho)$. The ratio $V^{(n+1)}(r)/V^{(n)}(r)$ between the $(n+1)$th and the nth derivatives is obviously independent of r and equal to $-1/\rho$ for this kind of potential. If the derivatives of two of these potentials are equal when the interatomic distances have the values r_{g1} and r_{g2}, respectively, i.e.,

$$V_1'(r_{g1}) = V_2'(r_{g2})$$

the ratio between the corresponding second derivatives will be

$$V_1''(r_{g1})/V_2''(r_{g2}) = [-V_1'(r_{g1})/\rho_1]/[-V_2'(r_{g2})/\rho_2] = \rho_2/\rho_1 \approx 1$$

The constants ρ are of rather uniform magnitude. Thus the two constants of immediate interest here, that for hydrogen–hydrogen and that for hydrogen–bromine interaction, have the values 0·3917 Å and 0·4379 Å, respectively.

REFERENCES

Badar, Y., Cooke, A. S., and Harris, M. M. (1965). *J. Chem. Soc.* 1412.
Bartell, L. S. (1960). *Tetrahedron Letters* No. 6, 13.
Bartell, L. S. (1961a). *J. Am. Chem. Soc.* 83, 3567.
Bartell, L. S. (1961b). *Iowa State J. Science* 36, 137.
Bastiansen, O. (1950). *Acta Chem. Scand.* 4, 926.
Bigeleisen, J., and Wolfsberg, M. (1958). *Advan. Chem. Phys.* 1, 15.
Brown, H. C., and McDonald, G. J. (1966). *J. Am. Chem. Soc.* 88, 2514.
Brown, H. C., Azzaro, M. E., Koelling, J. G., and McDonald, G. J. (1966). *J. Am. Chem. Soc.* 88, 2520.
Carter, R. E., and Dahlgren, L. (1969). *Acta Chem. Scand.* 23, 504.
Carter, R. E., and Dahlgren, L. (1970). *Acta Chem. Scand.* 24, 633.
Carter, R. E., and Junggren, E. (1968). *Acta Chem. Scand.* 22, 503.
Cooke, A. S., and Harris, M. M. (1963). *J. Chem. Soc.* 2365.
Clusius, K., and Weigand, K. (1940). *Z. physik. Chem.* B46, 1.
Grigor, A. F., and Steele, W. A. (1968). *J. Chem. Phys.* 48, 1032.
Halevi, E. A. (1963). *Progr. Phys. Org. Chem.* 1, 109.
Hampson, G. C., and Weissberger, A. (1936). *J. Am. Chem. Soc.* 58, 2111.
Heitner, C., and Leffek, K. T. (1966). *Can. J. Chem.* 44, 2567.
Horeau, A. (1961). *Tetrahedron Letters* No. 15, 506.
Horeau, A. (1962). *Tetrahedron Letters* No. 21, 965.
Horeau, A., and Nouaille, A. (1966). *Tetrahedron Letters* No. 33, 3953.
Horeau, A., Nouaille, A., and Mislow, K. (1965). *J. Am. Chem. Soc.* 87, 4957.
Howlett, K. E. (1960). *J. Chem. Soc.* 1055.
Kaplan, E. D., and Thornton, E. R. (1967). *J. Am. Chem. Soc.* 89, 6644.
Karabatsos, G. J., Sonnichsen, G. C., Papaioannou, C. G., Scheppele, S. E., and Shone, R. L. (1967). *J. Am. Chem. Soc.* 89, 463.
Karabatsos, G. J., and Papaioannou, C. G., (1968). *Tetrahedron Letters* No. 22, 2629.
Kreevoy, M. M., and Eyring, H. (1957). *J. Am. Chem. Soc.* 79, 5121.
Leffek, K. T., Llewellyn, J. A., and Robertson, R. E. (1960a). *Can. J. Chem.* 38, 1505.
Leffek, K. T., Llewellyn, J. A., and Robertson, R. E. (1960b). *J. Am. Chem. Soc.* 82, 6315.
Leffek, K. T., Llewellyn, J. A., and Robertson, R. E. (1960c). *Can. J. Chem.* 38, 2171.
Leffek, K. T., and Matheson, A. F. (1971). *Can. J. Chem.* 49, 439.
Melander, L., and Carter, R. E. (1964). *Acta Chem. Scand.* 18, 1138.
Mislow, K., Graeve, R., Gordon, A. J., and Wahl, G. H., Jr. (1964). *J. Am. Chem. Soc.* 86, 1733; preliminary communication: *Ibid.* 85, 1199 (1963).
Roussel, C., Chanon, M., and Metzger, J. (1971). *Tetrahedron Letters* No. 21, 1861.
Scott, R. A., and Scheraga, H. A. (1965). *J. Chem. Phys.* 42, 2209.
Shiner, V. J., Jr., and Humphrey, J. S., Jr. (1963). *J. Am. Chem. Soc.* 85, 2417.

Streitwieser, A., Jr., and Klein, H. S. (1964). *J. Am. Chem. Soc.* **86**, 5170.

Streitwieser, A., Jr., Jagow, R. H., Fahey, R. C., and Suzuki, S. (1958). *J. Am. Chem. Soc.* **80**, 2326.

Westheimer, F. H. (1947). *J. Chem. Phys.* **15**, 252.

Westheimer, F. H., and Mayer, J. E. (1946). *J. Chem. Phys.* **14**, 733.

Wilson, E. B., Jr., Decius, J. C., and Cross, P. C. (1955). "Molecular Vibrations. The Theory of Infrared and Raman Vibrational Spectra", p. 179. McGraw-Hill, New York, Toronto, London.

Wolfsberg, M., and Stern, M. J. (1964). *Pure Appl. Chem.* **8**, 225, 325.

THE REACTIVITY OF CARBONIUM IONS
TOWARDS CARBON MONOXIDE

H. HOGEVEEN*

Koninklijke/Shell-Laboratorium, Amsterdam. The Netherlands
(Shell Research N.V.)

I. Introduction	29
II. Reactivity and Stabilization of Tertiary Alkyl Cations	31
III. Reactivity and Stabilization of Secondary Alkyl Cations	34
A. Interconversion of Tertiary and Secondary Pentyloxocarbonium Ions	34
B. Rate Expressions for the Rearrangement-Carbonylation	37
C. Interconversion of Tertiary and Secondary Butyloxocarbonium Ions	40
D. Kinetic and Thermodynamic Control in the Reversible Carbonylation of the 2-Norbornyl Cation	41
IV. Interception of Unstable Cations by Carbon Monoxide	43
A. Primary Alkyl Cations	43
B. Vinyl Cations	45
V. Reactivity of Stabilized Carbonium Ions	46
VI. Some Remarks on the Mechanism of Carbonylation	50
VII. Conclusions	51
References	51

I. INTRODUCTION

THE reaction between carbonium ions and carbon monoxide affording oxocarbonium ions (acyl cations) is a key step in the well-known Koch reaction for making carboxylic acids from alkenes, carbon monoxide and water:

$$\text{Alkene} + H_2O + CO \xrightarrow{\ H^+\ } RCO.OH \tag{1}$$

The synthetic aspects of this acid-catalyzed process have been studied extensively for many years, and they have been reviewed recently by Falbe (1970). The kinetic and thermodynamic details of the various reaction steps in the overall process have been investigated in this laboratory during the last few years with special emphasis on the carbonylation step. The present article reflects the state of affairs in this respect.

The Koch reaction comprises the following reversible steps (2)–(5):

$$\text{Alkene} + H^+ \rightleftarrows R^+ \tag{2}$$
$$R^+ + CO \rightleftarrows RCO^+ \tag{3}$$
$$RCO^+ + H_2O \rightleftarrows RCO.OH_2^+ \tag{4}$$
$$RCO.OH_2^+ \rightleftarrows RCO.OH + H^+. \tag{5}$$

* Present address: Department of Organic Chemistry, The University, Zernikelaan Groningen, The Netherlands.

From a combination of data of two other reactions, the equilibrium constant for step (2) in the case of $R = t\text{-}C_4H_9$ was estimated to be $K = [t\text{-}C_4H_9{}^+]/[i\text{-}C_4H_8][H^+] = 10^{15 \cdot 8}$ litre mole^{-1} at 0°C in HF—SbF$_5$ solution (Hogeveen and Bickel, 1967). The rate constants of hydration and dehydration (equation (4)) have been determined in HF—BF$_3$ solution. In the case of $R = CH_3$ the former was found to be 2×10^3 litre mole^{-1} sec^{-1} (at 0°C) and the latter $9 \cdot 2 \times 10^{-4}$ sec^{-1} (at -13°C) (Hogeveen, 1967a, b). The deprotonation step in equilibrium (5) was shown to be diffusion-controlled in HF—BF$_3$ solution with HF$_2^-$ serving as the base (Hogeveen *et al.*, 1967).

All these data could be obtained by means of two techniques, namely n.m.r. spectroscopy and the use of superacid solvent systems (such as HF—BF$_3$, HF—SbF$_5$, FHSO$_3$—SbF$_5$, SbF$_5$—SO$_2$). As will become evident in this article, this is equally true for the data of the carbonylation and decarbonylation reactions (3). With less acidic systems the overall kinetics can, of course, be obtained but lack of knowledge concerning the concentrations of the intermediate ions prevents the determination of the rate constants of the individual steps.[1]

The two main reasons for studying the reversible reaction (3) were (a) to complete the picture of the Koch reaction in terms of quantitative information and (b) to set up a scale of reactivity towards a neutral nucleophile for carbonium ions of different structure. The first item is important from a practical point of view because there are reactions competing with the carbonylation step (3), which can be divided into *intra*molecular and *inter*molecular processes. *Rearrangement of the intermediate alkylcarbonium ion*, e.g.

$$CH_3.CH_2.CH\text{=}CH.CH_3$$

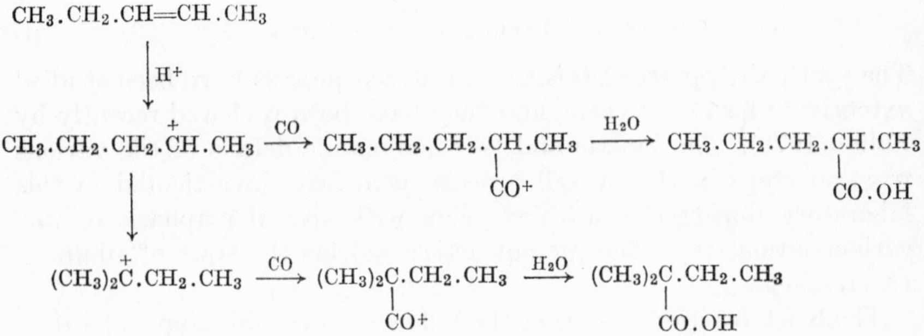

is an intramolecular process, and *dimerization of the intermediate alkylcarbonium ion*, e.g.

---

[1] An overall kinetic study on the Koch synthesis of succinic acid from acrylic acid and carbon monoxide in SO$_3$—H$_2$SO$_4$ has recently been reported (Sugita *et. al.*, 1970).

$(CH_3)_2C{=}CH_2$

$\downarrow H^+$

$(CH_3)_3C^+ \xrightarrow{\ CO\ } (CH_3)_3C\,.\,CO^+ \xrightarrow{\ H_2O\ } (CH_3)_3C\,.\,CO_2H$

$\downarrow {\scriptstyle (CH_3)_2C=CH_2}$

$(CH_3)_3C\,.\,CH_2\,.\,\underset{+}{C}(CH_3)_2 \xrightarrow{\ CO\ } (CH_3)_3C\,.\,CH_2\,.\,\underset{\underset{CO^+}{|}}{C}(CH_3)_2 \xrightarrow{\ H_2O\ } (CH_3)_3C\,.\,CH_2\,.\,\underset{\underset{CO\,.\,OH}{|}}{C}(CH_3)_2$

is an intermolecular process.

The type of carboxylic acid formed therefore depends on, amongst other things, the relative rates of carbonylation and of rearrangement or dimerization. The trialkylacetic acids are, from a technical point of view, the most valuable isomers (Vegter, 1970).

In previous studies (Hogeveen, 1970) the reactivity of long-lived carbonium ions towards molecular hydrogen has been investigated and interesting differences between secondary and tertiary alkyl cations have been observed. This reaction is too slow, however, to be extended to other types of carbonium ions. The reactivity of carbonium ions towards carbon monoxide is much higher (about six powers of ten) than towards molecular hydrogen, which enabled us to determine not only the rate of reaction (3) for some tertiary and secondary alkyl cations, but also the rate of carbonylation of more stabilized carbonium ions.

The experimental data for the carbonylation and decarbonylation have been obtained in super-acidic solutions with one or more of the following three techniques:

(a) measurement of the rate of carbon monoxide consumption or production in a conventional reactor. The pressure is kept constant by adding or letting off carbon monoxide at appropriate intervals. This technique is limited to rates corresponding to (pseudo) first-order rate constants of $k \leq 10^{-3}\ \text{sec}^{-1}$;

(b) measurements by n.m.r. spectroscopy of the change in concentration of reactant and/or product with time; the limitation of this technique is the same as that under (a);

(c) n.m.r. line-broadening measurements of relatively fast reactions in equilibrium, with $1 < k < 10^4\ \text{sec}^{-1}$.

II. REACTIVITY AND STABILIZATION OF TERTIARY ALKYL CATIONS

The reaction of t-butyl cation with carbon monoxide was found to be very much faster than that with hydrogen. Both this carbonylation and its reverse—the decarbonylation of the pivaloyl cation—are so rapid that the n.m.r. line-broadening technique could be applied in both the slow and the rapid exchange region in the temperature range of 10°C to 80°C (Hogeveen *et al.*, 1970):

$$\text{t-C}_4\text{H}_9^+ + \text{CO} \underset{k_D}{\overset{k_C}{\rightleftharpoons}} \text{t-C}_4\text{H}_9\text{CO}^+ \tag{6}$$

The observed line-broadening behaviour of the signals of the $\text{t-C}_4\text{H}_9^+$ ion (at 3·95 p.p.m.) and of the $\text{t-C}_4\text{H}_9\text{CO}^+$ ion (at 2·07 p.p.m.) is due to the *bimolecular–monomolecular* Koch equilibrium (6) and not to the conceivable *bimolecular–bimolecular* exchange equilibrium (7):

$$\text{t-C}_4\text{H}_9\text{CO}^+ + \text{t-C}_4\text{H}_9^+ \overset{k_E}{\rightleftharpoons} \text{t-C}_4\text{H}_9^+ + \text{t-C}_4\text{H}_9\text{CO}^+. \tag{7}$$

This was proved by showing that the reciprocal of the life-time of the pivaloyl ion, $\tau_{\text{RCO}^+}^{-1}$, is independent of the concentration of t-butyl ion. A mechanism based on equation (7) would have resulted in $\tau_{\text{RCO}^+}^{-1}$ being proportional to $[\text{t-C}_4\text{H}_9^+]$. The rate constant of decarbonylation of $\text{t-C}_4\text{H}_9\text{CO}^+$ in HF—SbF$_5$ (equimolar) and in FHSO$_3$—SbF$_5$ (equimolar) was determined to be $k_D = 10^{(13.0\pm0.6)} \times e^{(-15600\pm800)/RT}$ sec^{-1}. This extrapolates to a value of $3·2 \times 10^{-4}$ sec^{-1} at $-70°$C, which is in excellent agreement with that found from an experiment in which pivaloyl and isopropyl ions in SbF$_5$—SO$_2$ClF (1:2 v/v) solution gave t-butyl and isopropyloxocarbonium ions:

$$\text{t-C}_4\text{H}_9\text{CO}^+ \rightleftharpoons \text{t-C}_4\text{H}_9^+ + \text{CO}$$

$$(\text{CH}_3)_2\text{CH}^+ + \text{CO} \overset{\text{fast}}{\longrightarrow} (\text{CH}_3)_2\text{CHCO}^+$$

$$\overline{\text{t-C}_4\text{H}_9\text{CO}^+ + (\text{CH}_3)_2\text{CH}^+ \longrightarrow \text{t-C}_4\text{H}_9^+ + (\text{CH}_3)_2\text{CHCO}^+}$$

The rate constant of formation of isopropyloxocarbonium ions (irreversible at low temperatures, Section III, A) at $-70°$C was found to be $3·9 \times 10^{-4}$ sec^{-1}.

Calculation of the second-order rate constant of carbonylation, k_C, and the equilibrium constant, $K = [\text{t-C}_4\text{H}_9\text{CO}^+]/[\text{t-C}_4\text{H}_9^+][\text{CO}] = k_C/k_D$, requires knowledge of the concentration of CO. The constant α in Henry's law $P_{\text{CO}} = \alpha[\text{CO}]$ was determined to be 5·3 litre mole^{-1} atm in HF—SbF$_5$ (equimolar) and 53 litre mole^{-1} atm in FHSO$_3$—SbF$_5$ (equimolar) at 20°C. From the ratio $[\text{t-C}_4\text{H}_9\text{CO}^+]/[\text{t-C}_4\text{H}_9^+]$ at a known CO pressure, values for k_C and K were obtained. The data are listed in Table 1, which includes the values for the rate and equilibrium constants of two other tertiary alkyl cations, namely the t-pentyl and the t-adamantyl ions (Hogeveen *et al.*, 1970).

The equilibrium constant K is the same for $R^+ = t\text{-}C_4H_9^+$ and $t\text{-}C_5H_{11}^+$. As also the rate constants of carbonylation and decarbonylation are about equal for these two ions, it is concluded that both the thermodynamics and the kinetics of the carbonylation reaction are independent of the structure of R^+, if R^+ is an acyclic tertiary alkyl cation. This agrees with former findings (Brouwer, 1968) on the relative stabilities of such ions.

TABLE 1

Rate and Equilibrium Constants of $t\text{-}R^+ + CO \underset{k_D}{\overset{k_C}{\rightleftharpoons}} t\text{-}RCO^+$ at 20°C in FHSO$_3$—SbF$_5$ (equimolar)

R^+	$10^{-4} k_C$, litre mole^{-1} sec^{-1}	k_D, sec^{-1}	$K = k_C/k_D$, litre mole^{-1}
t-butyl	3	43	700
t-butyl	0·3[a]	43[a]	70[a]
t-pentyl	3[b]	43[b]	700
t-adamantyl	22	11	20,000

[a] In HF—SbF$_5$ (equimolar)
[b] Approximate values. Precise determinations over a temperature range to determine the activation parameters were not possible, owing to the occurrence of other reactions (Section III, A).

In discussing the effect of structure on the stabilization of alkyl cations on the basis of the carbonylation–decarbonylation equilibrium constants, it is assumed that—to a first approximation—the stabilization of the alkyloxocarbonium ions does not depend on the structure of the alkyl group. The stabilization of the positive charge in the alkyloxocarbonium ion is mainly due to the resonance $RC^+ = O \leftrightarrow RC \equiv O^+$, and the effect of R on this stabilization is only of minor importance. It has been shown by Brouwer (1968a) that even in the case of (tertiary) alkylcarbonium ions, which would be much more sensitive to variation of R attached to the electron-deficient centre, the stabilization is practically independent of the structure of the alkyl groups. Another argument is found in the fact that the equilibrium concentrations of isomeric alkyloxocarbonium ions differ by at most a factor of 2–3 from each other (Section III). Therefore, the value of K provides a quantitative measure of the stabilization of an alkyl cation. In the case of R = t-adamantyl this equilibrium constant is 30 times larger than when R = t-butyl or t-pentyl, which means that the non-planar t-adamantyl ion is $RT \ln 30 = 2\cdot1$ kcal mole^{-1} less stabilized than the acyclic ions. The smaller stabilization of the t-adamantyl ion has also been concluded on the basis of heat of formation (Arnett, 1970), solvolytic rate data (Raber *et al.*, 1970) and conformational-analysis calculations (Gleicher and Schleyer, 1967).

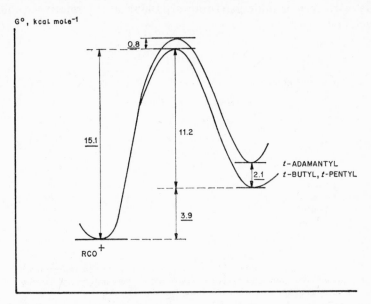

FIG. 1. Free-enthalpy diagram of the carbonylation–decarbonylation of tertiary alkyl cations at 20°C in FHSO$_3$—SbF$_5$ (concentrations expressed in mole litre^{-1}). Underlined numbers directly from experimental data.

This difference in stabilization is reflected in both the rate of carbonylation and decarbonylation (Table 1). The free-enthalpy profiles for the carbonylation of tertiary alkyl cations are shown in Fig. 1.

III. REACTIVITY AND STABILIZATION OF SECONDARY ALKYL CATIONS

A. *Interconversion of Tertiary and Secondary Pentyloxocarbonium Ions*

As mentioned in the Introduction, rearrangements of the intermediate alkyl cation in the Koch synthesis may compete with the carbonylation. Under the kinetically controlled conditions prevailing in the Koch synthesis of carboxylic acids, the rearrangements occur only from a less stable to a more stable carbonium ion, e.g. from a secondary to a tertiary ion. The reverse rearrangements—from a more stable to a less stable carbonium ion—can be observed if the carbonylation of the carbonium ion is performed under thermodynamically controlled conditions in $FHSO_3$—SbF_5 solution.

An example of thermodynamic control of oxocarbonium ion formation has been found (Hogeveen and Roobeek, 1970) in the dehydration of the pentylcarboxylic acids **1**, **2**, **3** or **4** in $FHSO_3$—SbF_5 at 40–60°.

$$
\begin{array}{ccc}
\underset{\underset{CO_2H}{|}}{C-\overset{\overset{C}{|}}{C}-C-C} \longrightarrow &
\left[\; \underset{\underset{CO^+}{|}}{C-\overset{\overset{C}{|}}{C}-C-C} \text{ and } C-\overset{\overset{C}{|}}{C}-\underset{\underset{CO^+}{|}}{C}-C \text{ and} \;\right] &
\longleftarrow \underset{\underset{CO_2H}{|}}{C-\overset{\overset{C}{|}}{C}-C-C} \\
(1) & (5) \qquad\qquad (6) & (2)
\end{array}
$$

$$
\begin{array}{ccc}
\underset{\underset{CO_2H}{|}}{C-C-C-C-C} \longrightarrow &
\left[\; \underset{\underset{CO^+}{|}}{C-C-C-C-C} \text{ and } C-C-\underset{\underset{CO^+}{|}}{C}-C-C \;\right] &
\longleftarrow \underset{\underset{CO_2H}{|}}{C-C-C-C-C} \\
(3) & (7) \qquad\qquad (8) & (4)
\end{array}
$$

Irrespective of the precursor acid the same mixture $(3:2:1:1)$ of ions **5**, **6**, **7** and **8** was eventually formed. For reasons of simplicity a $1:1:1:1$ ratio distribution will be assumed in the present discussion. The mechanism of the interconversion of ions **5–8** is as follows:

$$
\underset{\underset{CO^+}{|}}{C-\overset{\overset{C}{|}}{C}-C-C} \underset{CO}{\overset{-CO}{\rightleftharpoons}}
\underset{+}{C-\overset{\overset{C}{|}}{C}-C-C} \rightleftharpoons
C-\overset{\overset{C}{|}}{\underset{+}{C}}-C-C \underset{-CO}{\overset{CO}{\rightleftharpoons}}
\underset{\underset{CO^+}{|}}{C-\overset{\overset{C}{|}}{C}-C-C}
$$
$$
(5) \qquad\qquad (9) \qquad\qquad (10) \qquad\qquad (6)
$$

$$
\Updownarrow
$$

$$
\underset{\underset{CO^+}{|}}{C-C-C-C-C} \underset{CO}{\overset{-CO}{\rightleftharpoons}}
C-C-\overset{+}{C}-C-C \rightleftharpoons
\overset{+}{C}-C-C-C-C \underset{-CO}{\overset{CO}{\rightleftharpoons}}
\underset{\underset{CO^+}{|}}{C-C-C-C-C}
$$
$$
(7) \qquad\qquad (12) \qquad\qquad (11) \qquad\qquad (8)
$$

The rearrangements of the intermediate alkyl cations have been studied in great detail and reviewed recently by Brouwer and Hogeveen (1972) The scheme shows, besides the known decarbonylation of the tertiary pentyloxocarbonium ion (Section II), also decarbonylation of the secondary pentyloxocarbonium ions. The detailed free enthalpy (Gibbs free energy) diagram for the interconversion of ions **5** and **6** is given in Fig. 2. The values of $G°$ in this Figure are derived from the data in Fig. 1, the overall rate of conversion of **5** to **6** ($k = 1 \cdot 5 \times 10^{-3}$ sec^{-1} at 20°C; Hogeveen and Roobeek, 1970), the estimated free-enthalpy difference of 10 kcal mole^{-1} between **9** and **10** (Brouwer and Hogeveen, 1972; See also Section III, A), and the free-enthalpy barrier of about 2 kcal mole^{-1} for the 1,2-H shift in the step **10→9** (Brouwer and Hogeveen, 1972).

The salient points in this diagram are (i) the rate-determining step in the interconversion **5⇄6** is the bond-making (or bond-breaking) between the secondary C^{+} and CO; (ii) the rate of carbonylation of the secondary pentyl ion **10** (and presumably also of other secondary acyclic alkyl cations) in FHSO$_3$—SbF$_5$ has a free-enthalpy of activation of about

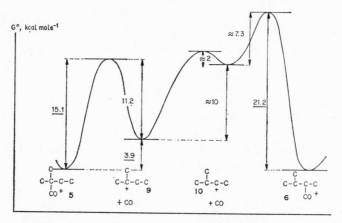

FIG. 2. Free-enthalpy diagram of the interconversion **5⇄6** at 20°C in FHSO$_3$—SbF$_5$ (concentrations expressed in mole litre^{-1}). Underlined numbers directly from experimental data.

7·3 kcal mole^{-1}, which means a rate constant of carbonylation of about 3×10^7 litre mole^{-1} sec^{-1} at 20°C.

The *tertiary-secondary* 1,2-H shift **9⇄10** is not rate-determining in the interconversion of **5** and **6**, but may become so in a conformationally fixed system. It has been found for the interconversion of tertiary and secondary adamantyloxocarbonium ions that $k < 10^{-4}$ sec^{-1} at 70°C (Hogeveen and Roobeek, 1971a) as compared with $k = 1 \cdot 5 \times 10^{-3}$ sec^{-1} at 20°C for the reaction **5⇄6**. The absence of interconversion between tertiary and secondary adamantyloxocarbonium ions is due to the circumstance that 1,2-H shifts do not occur in the tertiary adamantyl ion as a result of the effect of orbital orientation (Brouwer and Hogeveen, 1970; Schleyer *et al.*, 1970). That the secondary adamantyloxocarbonium ion can lose CO is demonstrated by the reaction with isopropyl cation in SbF$_5$—SO$_2$ClF solution at 0°C with formation of, for example, isopropyloxocarbonium ion (Hogeveen and Roobeek, 1971a).

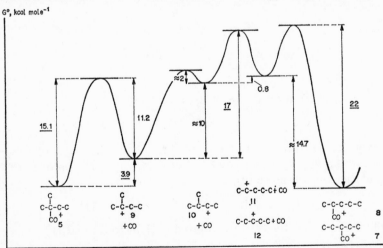

The isomerization of **5** to **7** and **8** involves a chain-branching t. pe rearrangement (**10⇌11**) (Brouwer and Oelderik; 1968) and has a free-enthalpy of activation of about 22 kcal mole^{-1}. This result, combined with the data of Fig. 2, the free-enthalpy of activation of 17 kcal mole^{-1} for the rearrangement **9→11** (Brouwer and Hogeveen, 1972), and an estimated difference in free-enthalpy of about 0·8 kcal mole^{-1} between **10** and **11** constitutes the basis for the free-enthalpy diagram in Fig. 3.

FIG. 3. Free-enthalpy diagram of the interconversion **5⇌7** and **8** at 20°C in FHSO$_3$—SbF$_5$ (concentrations expressed in mole litre^{-1}). The low barrier between ions **11** and **12** has been omitted. Underlined numbers directly from experimental data.

It is seen from Fig. 3 that the barrier of the chain-branching rearrangement $9 \rightleftarrows 11$ and 12 has about the same height as that of the carbonylation of 11 and 12, if $[CO] = 1$ mole litre^{-1}. Under the experimental conditions where $[CO]$ is about 10^{-2} mole litre^{-1}, the carbonylation (decarbonylation) step is rate-determining, however, and the transition state of highest free-enthalpy is therefore the same as in Fig. 2.

Before showing an example where the chain-branching rearrangement and not the carbonylation is rate-determining, the general rate equations for these processes will be discussed.

B. *Rate Expressions for the Rearrangement-Carbonylation*

The general scheme of the isomerization of alkyloxocarbonium ions is as follows:

$$R_1CO^+ \xrightleftharpoons[k_{-1}]{k_1} R_1^+ + CO \tag{7}$$

$$R_1^+ \xrightleftharpoons[k_{-2}]{k_2} R_2^+ \tag{8}$$

$$R_2^+ + CO \xrightleftharpoons[k_{-3}]{k_3} R_2CO^+ \tag{9}$$

$$R_1CO^+ \xrightleftharpoons[k_{-0}]{k_0} R_2CO^+ \tag{10}$$

The rate expression for the overall reaction (10) is

$$d[R_2CO^+]/dt = k_0[R_1CO^+] - k_{-0}[R_2CO^+] \tag{11}$$

To express the overall rate constants k_0 and k_{-0} in terms of the rate constants k_1, k_{-1}, k_2, k_{-2}, k_3 and k_{-3}, the rate of formation of R_2CO^+ is formulated, according to (9), as

$$d[R_2CO^+]/dt = k_3[R_2^+][CO] - k_{-3}[R_2CO^+] \tag{12}$$

Application of the steady-state approximations to R_2^+ and R_1^+ gives

$$d[R_1^+]/dt = k_1[R_1CO^+] - k_{-1}[R_1^+][CO] - k_2[R_1^+] + k_{-2}[R_2^+] = 0$$
$$d[R_2^+]/dt = k_3[R_2CO^+] - k_3[R_2^+][CO] - k_{-2}[R_2^+] + k_2[R_1^+] = 0,$$

from which it follows that

$$[R_2^+] = \frac{k_1 k_2 [R_1CO^+] + k_2 k_{-3}[R_2CO^+] + k_{-1}k_{-3}[R_2CO^+][CO]}{k_2 k_3[CO] + k_{-1}k_3[CO]^2 + k_{-1}k_{-2}[CO]} \tag{13}$$

Substitution of equation (13) in equation (12) gives

$$d[R_2CO^+]/dt = \frac{k_1 k_2 k_3 [R_1CO^+]}{k_{-1}k_{-2} + k_2 k_3 + k_{-1}k_3[CO]} - \frac{k_{-1}k_{-2}k_{-3}[R_2CO^+]}{k_{-1}k_{-2} + k_2 k_3 + k_{-1}k_3[CO]} \tag{14}$$

Comparison of equations (11) and (14) shows that

$$k_0 = \frac{k_1 k_2 k_3}{k_{-1}k_{-2} + k_2 k_3 + k_{-1}k_3[CO]} \tag{15}$$

$$k_{-0} = \frac{k_{-1}k_{-2}k_{-3}}{k_{-1}k_{-2} + k_2 k_3 + k_{-1}k_3[CO]} \tag{15'}$$

In the case of equal thermodynamic stabilities of R_1CO^+ and R_2CO^+ ions, $k_0 = k_{-0}$.

If R_1^+ is a tertiary and R_2^+ a secondary alkyl cation, equation (15) can be simplified to

$$k_0 = \frac{k_1 k_2 k_3}{k_{-1}k_{-2} + k_{-1}k_3[CO]} \tag{16}$$

because $k_{-2} \gg k_2$ and $k_{-1} < k_3$. Equation (16) is further simplified to (17)

$$k_0 = \frac{k_1 k_2 k_3}{k_{-1}k_{-2}}, \tag{17}$$

if $k_3[CO] \ll k_{-2}$, or to (18)

$$k_0 = \frac{k_1 k_2}{k_{-1}[CO]}, \tag{18}$$

if $k_3[CO] \gg k_{-2}$.

Equation (17) pertains to the situations described in Figs. 2 and 3, in which the carbonylation step k_3—or in the reverse direction, the decarbonylation step k_{-3}—is rate-limiting. It should be remarked that in the case of Fig. 3 the rearrangement step k_2 might become rate-limiting if the reaction were carried out at a very high CO pressure, that is, the condition for (18) rather than for (17) would then be fulfilled.

An example to which equation (18) applies at a CO pressure of one atmosphere has been found (Hogeveen and Roobeek, 1970) in the interconversion of tertiary and secondary butyloxocarbonium ions (Section III, C).

C. *Interconversion of Tertiary and Secondary Butyloxocarbonium Ions*

Tertiary and secondary butyloxocarbonium ions interconvert according to

$$
\underset{\substack{|\\ \text{C}}}{\overset{\substack{\text{C}\\|}}{\text{C--C--CO}^+}}
\underset{\text{CO}}{\overset{-\text{CO}}{\rightleftharpoons}}
\underset{\substack{|\\ \text{C}}}{\overset{\substack{\text{C}\\|}}{\text{C--C}^+}}
\rightleftharpoons
\text{C--C--}\overset{+}{\text{C}}\text{--C}
\underset{-\text{CO}}{\overset{\text{CO}}{\rightleftharpoons}}
\underset{\substack{|\\ \text{CO}^+}}{\text{C--C--C--C}}
$$

but much more slowly than the homologous pentyloxocarbonium ions. The reasons for this is the low rate of rearrangement of the intermediate butyl cation (Brouwer and Oelderik, 1968; Brouwer, 1968b; Saunders *et al.*, 1968). Consequently, rate equation (18) is applicable to this case. Extrapolation of the rate constants at 80–110°C gives a value for k_0 at 20°C of 1.2×10^{-8} sec^{-1} at a CO pressure of 1 atm, corresponding to [CO] = 0.019 mole litre^{-1} (Hogeveen *et al.*, 1970). Equation (18) shows the overall rate constant k_0 to be inversely proportional to [CO]—in contrast to equation (17)—and therefore k_0 at [CO] = 1 mole litre^{-1} is

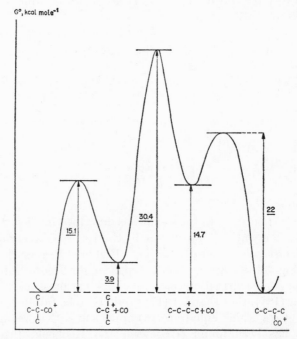

Fig. 4. Free-enthalpy diagram of the interconversion of tertiary and secondary butyloxocarbonium ions at 20°C in FHSO₃—SbF₅ (concentrations expressed in mole litre^{-1}). Underlined numbers directly from experimental data.

equal to $1.2 \times 10^{-8} \times 0.019 = 2.3 \times 10^{-10}$ sec^{-1} and $\Delta G^{\ddagger} = 30.4$ kcal mole^{-1}. The free-enthalpy diagram in Fig. 4 has been constructed on the basis of these data and those in Fig. 3.

From Fig. 4 it is seen that the free-enthalpy of activation for the rearrangement of tertiary butyl to secondary butyl cation is $30.4 - 3.9 = 26.5$ kcal mole^{-1}. As the reverse rearrangement has been found by direct observation to have $\Delta G^{\ddagger} \approx 17$–$18$ kcal mole^{-1} (Saunders et al., 1968), it follows that the difference in stabilization between tertiary and secondary butyl cations is indeed 9 ± 1 kcal mole^{-1}. This value is in excellent agreement with a previous "experimental" value of 10 ± 1 kcal mole^{-1} (Brouwer and Hogeveen, 1972).

The difference in behaviour between pentyl and butyl cation systems (Figs. 3 and 4) has also been encountered in trapping experiments with carbonium ions, primarily formed from alkanes and SbF$_5$, by CO (Hogeveen and Roobeek, 1972). In the case of n-butane the secondary butyloxocarbonium ion is the main product, whereas in the case of n-pentane only the tertiary pentyloxocarbonium ion is found.

D. Kinetic and Thermodynamic Control in the Reversible Carbonylation of the 2-Norbornyl Cation

The stable 2-norbornyl cation has recently been shown to be a non-classical, unusually stabilized species. Olah et al. (1970) proved spectroscopically that this ion is a corner-protonated nortricyclene with a pentavalent carbon atom. The value for the carbonylation–decarbonylation equilibrium constant $K\, (= k_C^{exo}/k_D^{exo})$ of the 2-norbornyl ion illustrates quantitatively its high stability (Hogeveen et al., 1970).

$$\ce{->[k_D^{exo}][k_C^{exo}]} \quad + \text{ CO} \quad \ce{<-[k_D^{endo}][k_C^{endo}]} \tag{19}$$

The value for K was determined to be 10^4 litre mole^{-1} at 20°C, which is of the same order of magnitude as those for tertiary alkyl cations $((0.07 - 2) \times 10^4$ litre mole^{-1}, Section II) and dramatically different from those for secondary alkyl cations (about 10^{10} litre mole^{-1}, calculated from Figs. 2 and 3). These data show that the 2-norbornyl ion is only 1.6 kcal mole^{-1} less stabilized than, for example, the tertiary butyl cation and about 8 kcal mole^{-1} more stabilized than secondary alkyl cations. Another thermodynamic argument for the high stability of the 2-norbornyl ion in solution is found in the work of Arnett and Larsen (1968) on heats of formation of carbonium ions. (For the gas-phase stability see Kaplan *et al.*, 1970.)

Although carbonylation of the 2-norbornyl ion at or below room temperature leads to exclusive formation of the 2-*exo*-norbornyloxocarbonium ion, reactions at higher temperatures have shown that the 2-*endo*-norbornyloxocarbonium ion is just as stable as the *exo*-isomer (Hogeveen and Roobeek, 1969). This means that at low temperatures the carbonylation is *kinetically* controlled, and at high temperatures *thermodynamically* controlled. The detailed free-enthalpy diagram in

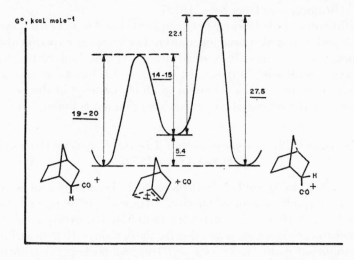

Fig. 5. Free-enthalpy diagram of the interconverson of 2-*endo*- and 2-*exo*-norbornyloxocarbonium ions via the 2-norbornyl ion at 20°C in FHSO₃—SbF₅ (concentrations expressed in mole litre⁻¹). Underlined numbers directly from experimental data.

Fig. 5 for the equilibrium (19) is based on the rates of interconversion of the 2-*exo*- and 2-*endo*-norbornyloxocarbonium ions, the above-mentioned value of K for the 2-*exo*-norbornyloxocarbonium ion, and the limits of the rates of formation and decarbonylation of the latter ion (Hogeveen and Roobeek, 1969).

A few points of interest in Fig. 5 are the following.

(a) The free-enthalpy of activation in the formation of 2-*exo*-norbornyloxocarbonium ion is about 8 kcal mole⁻¹ smaller than that of 2-*endo*-norbornyloxocarbonium ion. This shows that in the reaction of carbon monoxide with the corner-protonated cyclopropane ring of the 2-norbornyl ion an *inversion* mechanism (A) is strongly preferred ($\approx 10^6 : 1$) to a *retention* mechanism (B).

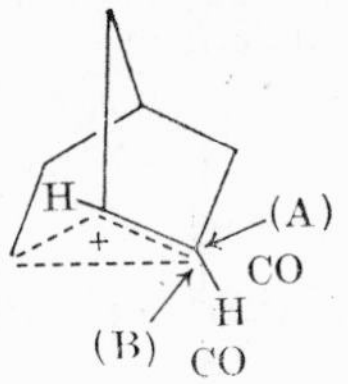

(b) Although the stabilization of the 2-norbornyl ion is $1 \cdot 6$ kcal mole^{-1} smaller than that of the t-butyl ion, the ΔG^{+} of carbonylation is about 3 kcal mole^{-1} larger for the former ion. The delocalization of the positive charge in the protonated cyclopropane ring of the 2-norbornyl ion is thought to be mainly responsible for this higher barrier; steric hindrance in the transition state between carbon monoxide and the 3-*exo* and 7-*syn* hydrogen atoms may be an additional factor[2] (Hogeveen and Roobeek, 1969; Hogeveen *et al.*, 1970).

(c) The diagram in Fig. 5 is similar to those of the solvolysis of 2-*endo*- and 2-*exo*-norbornyl derivatives ("Goering-Schewene" diagram; Goering and Schewene, 1965; Brown and Rei, 1968) the difference being that in solvolysis work the absolute value of the free-enthalpy of activation for the reaction of the norbornyl ion with the nucleophile is unknown, whereas in the carbonylation–decarbonylation system the complete free-enthalpy picture can be given.

IV. INTERCEPTION OF UNSTABLE CATIONS BY CARBON MONOXIDE

A. *Primary Alkyl Cations*

In Sections II and III it was shown that secondary and tertiary alkyl cations can be formed by decarbonylation of the corresponding oxo-carbonium ions. This has been found impossible in the case of primary alkyl cations (Hogeveen and Roobeek, 1970): the oxocarbonium ions **13** and **14** were unchanged after one hour at 100°C ($k < 1 \cdot 3 \times 10^{-5}$ sec^{-1}), whereas ion **15** is fragmented by a β-fission under these circumstances:

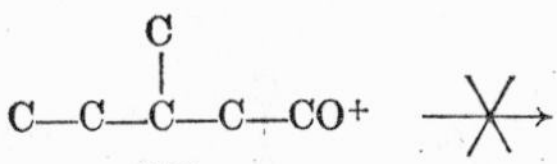

---

[2] This is unlikely to be a dominant factor. Steric effects have been observed in the Koch-Haaf synthesis of carboxylic acids (Pincock *et al.*, 1959; Stork and Bersohn, 1960; Peters and Van Bekkum, 1971), but these are ascribed to the steric requirements of the protonated carboxyl group ($—COOH_2^+$) and not to those of the oxocarbonium group ($—CO^+$). In FHSO$_3$—SbF$_5$, the product is the alkyloxocarbonium ion.

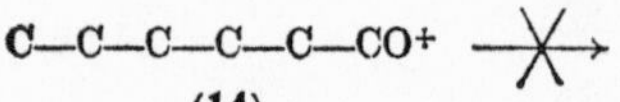

$$\text{C—C—C—C—C—CO}^+ \quad \overset{\times}{\longrightarrow}$$

(14)

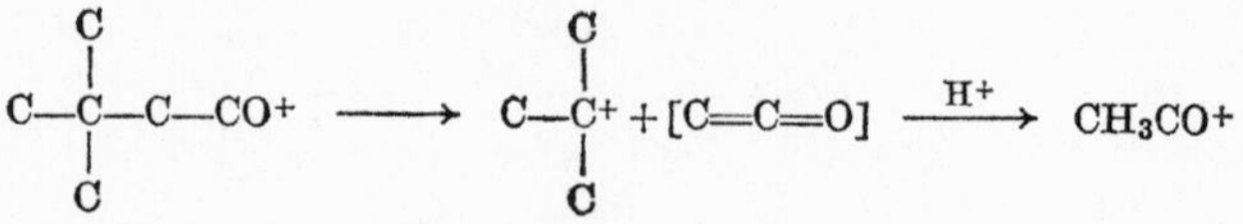

$$\underset{\overset{|}{\text{C}}}{\overset{\overset{\text{C}}{|}}{\text{C—C—C—CO}^+}} \longrightarrow \underset{\overset{|}{\text{C}}}{\overset{\overset{\text{C}}{|}}{\text{C—C}^+}} + [\text{C}{=}\text{C}{=}\text{O}] \xrightarrow{\text{H}^+} \text{CH}_3\text{CO}^+$$

(15)

The simplest primary alkyl cations, CH_3^+ and $C_2H_5^+$, are formed from methane and ethane, respectively, by SbF_5—$FHSO_3$ (Olah and Schlosberg, 1968; Olah *et al.*, 1969) and by SbF_5 (Lukas and Kramer, 1971). In these cases, intermolecular electrophilic substitution of these ions at the precursor alkanes leads to oligocondensation products, e.g. tertiary butyl and hexyl ions. In the presence of carbon monoxide it has been found possible to intercept the intermediate CH_3^+ and $C_2H_5^+$ quantitatively as oxocarbonium ions (Hogeveen *et al.*, 1969; Hogeveen and Roobeek, 1972). The competition between the reactions of the ethyl cation with ethane and carbon monoxide, respectively, is illustrated by the following equations:

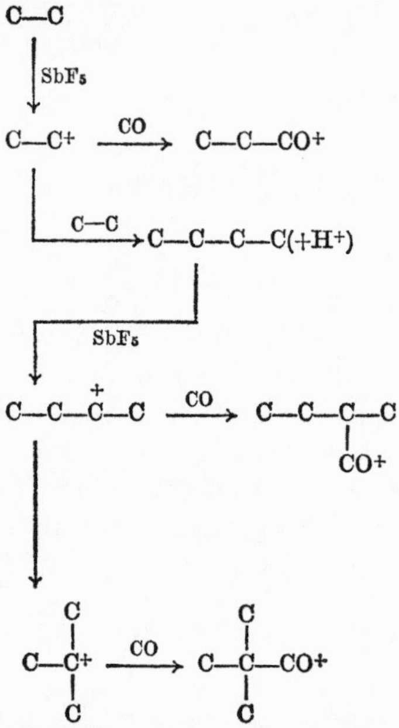

$$\text{C—C}$$
$$\downarrow \text{SbF}_5$$
$$\text{C—C}^+ \xrightarrow{\text{CO}} \text{C—C—CO}^+ \tag{20}$$
$$\downarrow \overset{\text{C—C}}{\longrightarrow} \text{C—C—C—C}(+\text{H}^+) \tag{21}$$

$$\downarrow \text{SbF}_5$$
$$\text{C—C—}\overset{+}{\text{C}}\text{—C} \xrightarrow{\text{CO}} \underset{\overset{|}{\text{CO}^+}}{\text{C—C—C—C}}$$

$$\downarrow$$

$$\underset{\overset{|}{\text{C}}}{\overset{\overset{\text{C}}{|}}{\text{C—C}^+}} \xrightarrow{\text{CO}} \underset{\overset{|}{\text{C}}}{\overset{\overset{\text{C}}{|}}{\text{C—C—CO}^+}}$$

From the product distribution at various CO—C_2H_6 ratios it was concluded that the dimerization step (21) has a rate constant of the same order of magnitude as that of the carbonylation step (20).

Although a value for the rate constant of carbonylation of primary alkyl cations has so far not been obtained experimentally, it can be shown that the reaction must be diffusion-controlled. The difference in stabilization between secondary and tertiary alkyl cations in solution (9 ± 1 kcal mole^{-1}; Section III, C) shows up as a difference in rate of carbonylation of a factor of 10^3 (Section III, A). As the difference in stabilization between primary and secondary alkyl cations is certainly larger than that between secondary and tertiary alkyl cations (various estimates have been summarized by Brouwer and Hogeveen, 1972), the rate constant of carbonylation of primary alkyl cations—and even more so for the methyl cation—will exceed that of secondary alkyl cations by more than a factor of 10^3, so that $k > 10^{10-11}$ litre mole^{-1} sec^{-1}, which means that the reaction is diffusion controlled.

It should be emphasized that, in contrast to the cases of secondary alkyl cations (Olah et al., 1964; Saunders et al., 1968) and tertiary alkyl cations (Olah et al., 1964; Brouwer and Mackor, 1964), the evidence for the existence of primary alkyl cations as distinct species has been only in-direct, because they have escaped direct spectroscopic observation so far.

B. Vinyl Cations

Evidence for the occurrence of vinyl cations as short-lived inter-mediates in solvolysis and other reactions has accumulated in the last few years (reviewed by Hanack, 1970, by Richey and Richey, 1970, and by Modena and Tonellato, 1971), but they have not been observed spectroscopically. It has been shown possible to intercept some vinyl cations—prepared in a system of extremely low nucleophilicity ($FHSO_3$—SbF_5 1:1–1:10) by protonation of propyne and 2-butyne—by carbon monoxide (Hogeveen and Roobeek, 1971b). The oxocarbo-nium ions formed in these cases are shown in the following scheme:

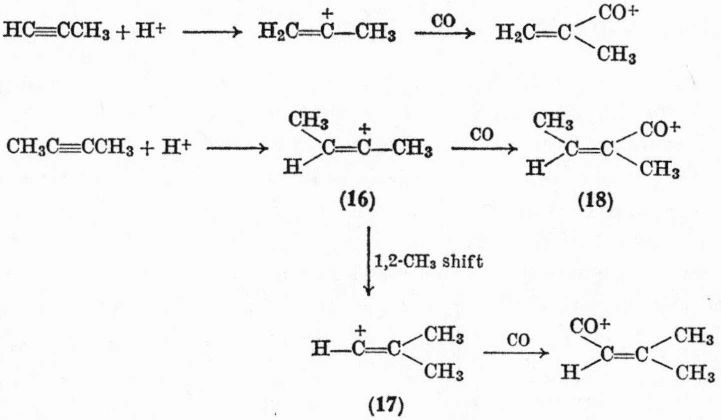

$$HC{\equiv}CCH_3 + H^+ \longrightarrow H_2C{=}\overset{+}{C}{-}CH_3 \xrightarrow{\ CO\ } H_2C{=}C\overset{\diagup CO^+}{\diagdown CH_3}$$

$$CH_3C{\equiv}CCH_3 + H^+ \longrightarrow \overset{CH_3}{\underset{H}{>}}C{=}\overset{+}{C}{-}CH_3 \xrightarrow{\ CO\ } \overset{CH_3}{\underset{H}{>}}C{=}C\overset{\diagup CO^+}{\diagdown CH_3}$$

(16)   (18)

1,2-CH$_3$ shift

$$H{-}\overset{+}{C}{=}C\overset{\diagup CH_3}{\diagdown CH_3} \xrightarrow{\ CO\ } \overset{CO^+}{\underset{H}{>}}C{=}C\overset{\diagup CH_3}{\diagdown CH_3}$$

(17)

In contrast to the results of the reaction of tertiary and secondary alkyl cations with carbon monoxide (Figs. 1–5), which were obtained under thermodynamically controlled conditions, the results of the carbonylation with the vinyl cations were obtained under kinetically controlled conditions. This presents a difficulty in explaining the occurrence of the 1,2-CH_3 shift in the reaction **16**→**17**, because it involves a strong increase in energy. The exclusive formation of the Z-stereoisomer **18** on carbonylation of the 1,2-dimethylvinyl cation **16** is remarkable,[3] but does not allow an unambiguous conclusion about the detailed structure—linear **19** or bent **20**—of the vinyl cation. A non-classical structure **21** can be disregarded, however, because the attack

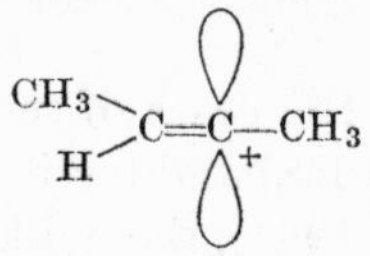

(19) 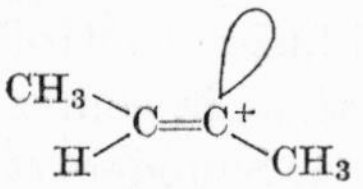   **(20)** 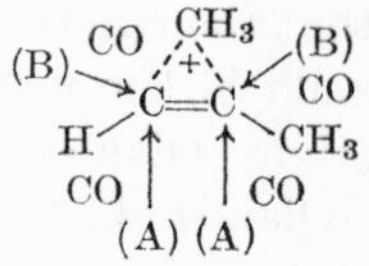   **(21)**

of CO on **21** according to pathway (A)—which is the kinetically preferred pathway in the protonated cyclopropane structure of the norbornyl cation, Section III, D—does not lead to the observed Z-isomer **18** but to the E-stereoisomer. Moreover, carbonylation of a nonclassical structure such as **21** for the methylvinyl cation (from propyne) would lead to the (Z)-3-methylpropenoyl ion rather than to the observed 2-methylpropenoyl ion (Hogeveen and Roobeek, 1971b).

Kinetic data on the carbonylation of vinyl cations have not been obtained so far, but it is likely to be a diffusion-controlled reaction as in the case of primary alkyl cations (Section IV, A).

V. REACTIVITY OF STABILIZED CARBONIUM IONS

In Sections II and III the quantitative aspects have been summarized of the reversible carbonylation of secondary and tertiary alkyl cations as studied under thermodynamically controlled conditions. In Section IV the results have been reviewed of the irreversible carbonylation of the much less stable primary alkyl and vinyl cations as studied under kinetically controlled conditions. No kinetic details had been obtained in the latter case owing to the short-lived character of the ions.

This section deals with the kinetics of the carbonylation of some stabilized cations (cyclopropylcarbonium ions, allylic and homoallylic

[3] In the metal–carbonyl catalysed hydrocarboxylation of alkynes ("Reppe reaction") nearly exclusive cis-addition of H—COOH is found (Ohashi et al., 1952).

cations, and aromatic cations) studied under kinetically controlled conditions (Hogeveen and Gaasbeek, 1970).

The kinetic results of these reactions are summarized in Table 2, in which the data for alkyl cations are included for comparison.

It should be noted that the kinetics were first-order over at least three half-lives (with the exception of the dicyclopropylcarbonium ion), but the reaction products were not well defined in some cases—probably due to relatively fast consecutive reactions of the unsaturated oxocarbonium ions formed. In the case of the oxocarbonium ions formed from the allyl cations a novel quantitative cyclization to give cyclopentenone derivatives was observed (Hogeveen and Gaasbeek, 1970):

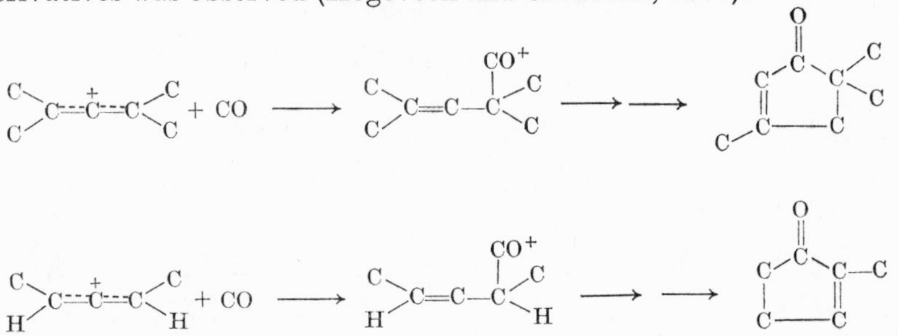

For the determination of stabilizations of carbonium ions the equilibrium constants of carbonylation–decarbonylation have been used in previous Sections. For the ions discussed in this Section, however, the rate constants of decarbonylation are not known and, therefore, the rate constants of carbonylation will be used as a criterion for such stabilizations. This kinetic criterion is a useful indicator if there are no significant steric factors in the carbonylation step and if this step is indeed rate-determining in the overall process (Hogeveen and Gaasbeek, 1970). The following rate constants in Table 2 are of particular importance.

(a) The rate of carbonylation of the 7-norbornadienyl cation is about 10^5 times smaller than the rates for cyclopropylcarbonium ions. This low reactivity of the 7-norbornadienyl cation is therefore not consistent with a description of this ion in terms of equilibrating structures (22) (Brown et al., 1965)—which are cyclopropyl carbonium ions—but in line with an additional non-classical stabilization (23) (Brookhart et al., 1967).

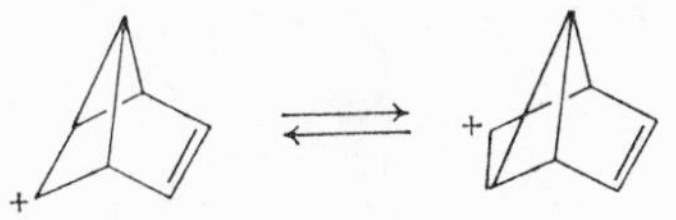

(22) 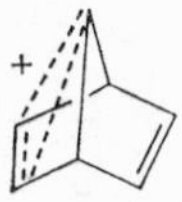   (23)

TABLE 2

Rate Constants of Carbonylation of Carbonium Ions: $R^+ + CO \rightarrow RCO^+$. $[R^+] = 0.3\text{–}0.7$ mole litre^{-1}. $P_{CO} = 15$ atm

R^+	Solvent		Temperature °C	$k'_C \cdot 10^3$ sec^{-1}	[CO] mole litre^{-1}	$k_C \cdot 10^3$ litre mole^{-1} sec^{-1}
n-$C_nH_{2n+1}^+$						$\sim 10^{13\text{–}14}$ [a]
s-$C_nH_{2n+1}^+$	FHSO$_3$—SbF$_5$	(1:1 m/m)	20			3×10^{10} [b]
t-$C_nH_{2n+1}^+$	HF/SbF$_5$	(1:1 m/m)	20			3×10^{6} [c]
	FHSO$_3$/SbF$_5$	(1:1 m/m)	20			3×10^{7} [c]
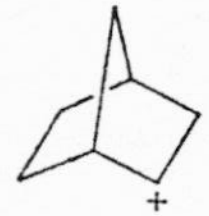	FHSO$_3$—SbF$_5$	(1:1 m/m)	20			$<10^5$ [e]
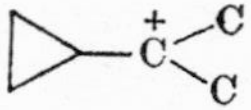	SbF$_5$—SO$_2$ClF	(1:2 v/v)	−68	1·6	0·45	3·6
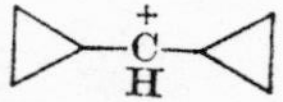	FHSO$_3$—SbF$_5$—SO$_2$	(1:1:3 m/m)	−68			<3 [d]
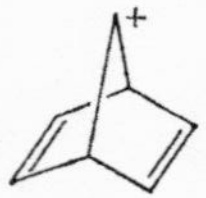	FHSO$_3$		1	1·1	0·25	4·4
	HF		−2	1·2	0·44	2·7
	HF		−15	0·26	0·56	0·46
	HF		−22	0·065	0·60	0·11
	HF		−68			2×10^{-5} [e]
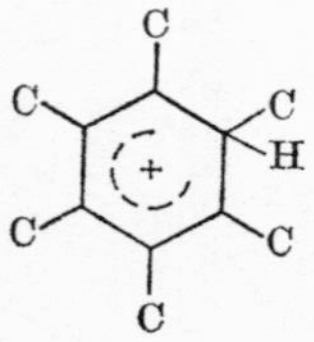	HF—SbF$_5$	(14:1 v/v)	20			$<10^{-2}$ [f]

Ion	Solvent	Temp.	k'_c	k_c		
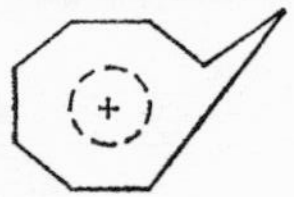 (cycloheptatrienyl cation)	HF	20				$<10^{-2\ f}$
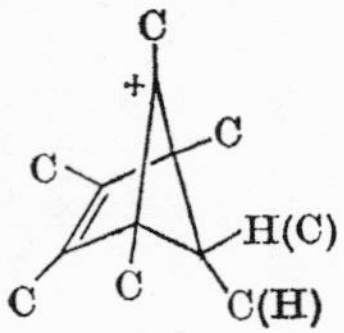 (bicyclic cation, H(C), C(H))	HF	−23				$<10^{-2\ f}$
CH_2^+–(phenyl)	SbF$_5$—SO$_2$ClF (1:2 v/v)	−67	$0\cdot55^{\,g}$	$0\cdot03$		$18\cdot4$
C–C$^+$–C–C	HF—SbF$_5$ (8:1 v/v)	0	$0\cdot02$	$0\cdot5^{\,h}$		$0\cdot04$
	FHSO$_3$—SbF$_5$ (1:1 m/m)	0	$0\cdot050$	$0\cdot4^{\,h}$		$0\cdot13$
C–C–C$^+$–C–C	FHSO$_3$—SbF$_5$ (1:1 m/m)	0	11	$0\cdot4^{\,h}$		30

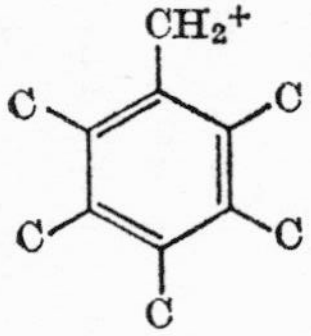

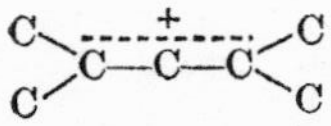

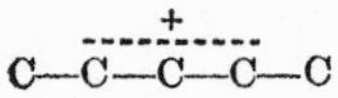

k'_c pseudo first-order rate constant.

k_c second-order rate constant.

a Extrapolated value; Section IV, A.

b Calculated from Fig. 2; Section III, A.

c Determined by n.m.r. line broadening; Sections II, III, D.

a No kinetics could be determined, because much more than an equivalent amount of CO was taken up.

e Calculated rate constant.

f No reaction after 6 hours.

g Carried out at 1 atm CO pressure.

h Approximate concentrations.

(b) The rates of carbonylation of the 7-norbornadienyl ion on the one hand and the 1,3-dimethyl- and 1,1,3,3-tetramethylallyl ions on the other hand permit us to compare the non-classical, homoconjugative stabilization in the former ion with the classical conjugative stabilization in the latter ions. It is anticipated that the stabilization of the former ion is about the same as that of the 1,1,3-trimethylallyl ion.

(c) The hexamethylbenzenium ion, the homotropylium ion and the hexamethylbicyclo [2,1,1] hexenyl ions are too strongly stabilized to react with carbon monoxide under the given conditions. The high reactivity of the pentamethylbenzyl ion towards carbon monoxide at low temperatures concurs with the reported reactivity of this ion towards molecular hydrogen and alkanes (Buck et al., 1964).

VI. Some Remarks on the Mechanism of Carbonylation

Although the addition of carbon monoxide to a carbonium ion R^+ leads to a product $R—C^+{=}O$, with a new carbon–carbon bond, one may raise the question whether this occurs by direct alkylation at carbon:

$$R^+ + CO \longrightarrow R—\overset{+}{C}{=}O$$

or by alkylation at oxygen, followed by a 1,2-alkyl shift

$$R^+ + CO \longrightarrow R—O\overset{+}{=}C$$

$$R—O\overset{+}{=}C \longrightarrow R—\overset{+}{C}{=}O$$

The intermediate in the second mechanism is identical with that postulated by Skell and Starer (1959) in the formation of carbonium ions from alkoxide and carbenes.

Although the first mechanism is intuitively, and for reasons of simplicity, preferred to the second one, there is no experimental evidence for excluding the latter. Theoretical evidence for preferring the former mechanism stems from a study by Jansen and Ros (1969), who performed non-empirical calculations on several configurations of a model system, viz. protonated carbon monoxide.[4] They found that the linear configuration $H—\overset{+}{C}{=}O$ is 33 kcal mole^{-1} more stable than the linear configuration $H—\overset{+}{O}{=}C$, and that non-linear arrangements are even less stable than the latter. An approach of the H^+ perpendicular to the C—O bond does not lead to a real energy minimum (saddle point), so

[4] This is a frequently postulated species (Olah and Kuhn, 1964; Ciuhandu and Dumitreanu, 1970) which so far has escaped spectroscopic detection (Hogeveen et al., 1967).

that the most likely reaction pathway is along the C—O axis. These calculations concur with the fact that the methyloxocarbonium ion has a linear structure in the crystalline state (Boer, 1968).

VII. Conclusions

The study of the carbonylation of carbonium ions, as summarized in this Review, has afforded valuable information on a number of problems.

First, the rates of carbonylation of secondary and tertiary alkyl carbonium ions can now be compared quantitatively with the known rates of competing intramolecular rearrangements of these ions. The product distribution in the Koch synthesis of carboxylic acids depends, amongst other things, on these relative rates.

Secondly, the stabilization of alkylcarbonium ions can be conveniently determined by measuring the equilibrium constants of the carbonylation–decarbonylation reactions. For some cases the rates of carbonylation are used as a kinetic criterion for stabilization.

Thirdly, trapping of highly unstable carbonium ions, such as primary alkyl and vinyl ions, by carbon monoxide has been shown possible, so that their existence as distinct species with a finite lifetime has gained in plausibility.

Finally, it should be noted that the range of reactivities studied is very large indeed, covering at least fifteen powers of ten between highly unstable and highly stabilized ions.

References

Arnett, E. M. (1970). Quoted by P. Vogel, M. Saunders, W. Thielecke and P. v. R. Schleyer.
Arnett, E. M., and Larsen, J. W. (1968) in "Carbonium Ions" (G. A. Olah and P. v. R. Schleyer, eds.), Vol. 1, p. 441, Interscience Publishers, New York.
Boer, F. P. (1968). J. Am. Chem. Soc. **90**, 6706.
Brookhart, M., Lustgarten, R. K., and Winstein, S. (1967). J. Am. Chem. Soc. **89**, 6352.
Brouwer, D. M. (1968a). Rec. Trav. Chim. **87**, 210.
Brouwer, D. M. (1968b). Rec. Trav. Chim. **87**, 1435.
Brouwer, D. M., and Hogeveen, H. (1970). Rec. Trav. Chim. **89**, 211.
Brouwer, D. M., and Hogeveen, H. (1972) in "Progress in Physical Organic Chemistry" (R. W. Taft and A. Streitwieser, Jr., eds.), Vol. 9, p. 179, Interscience Publishers, New York.
Brouwer, D. M., and Mackor, E. L. (1964). Proc. Chem. Soc. 147.
Brouwer, D. M., and Oelderik, J. M. (1968). Rec. Trav. Chim. **87**, 721.
Brown, H. C., and Rei, M. H. (1968). J. Am. Chem. Soc. **90**, 6216.
Brown, H. C., Morgan, K. J., and Chloupek, F. J., (1965). J. Am. Chem. Soc. **87**, 2137.
Buck, H. M., v. d. Sluys-van der Vlugt, M. J., Dekkers, H. P. J. M., Brongersma, H. H., and Oosterhoff, L. J. (1964). Tetrahedron Letters, 2987.

Ciuhandu, G., and Dumitreanu, A. (1970). *Liebigs Ann.* 742, 98.

Falbe, J. (1970). "Carbon Monoxide in Organic Synthesis", Chapter III, Springer-Verlag, Berlin.

Gleicher, G. J., and Schleyer, P. v. R. (1967). *J. Am. Chem. Soc.* 89, 582.

Goering, H. L., and Schewene, C. B. (1965). *J. Am. Chem. Soc.* 87, 3516.

Hanack, M. (1970). *Accounts Chem. Res.* 3, 209.

Hogeveen, H. (1967a). *Rec. Trav. Chim.* 86, 289.

Hogeveen, H. (1967b). *Rec. Trav. Chim.* 86, 809.

Hogeveen, H. (1970). *Rec. Trav. Chim.* 89, 74 and previous papers.

Hogeveen, H., and Bickel, A. F. (1967). *Rec. Trav. Chim.* 86, 1313.

Hogeveen, H., and Gaasbeek, C. J. (1970). *Rec. Trav. Chim.* 89, 395.

Hogeveen, H., and Roobeek, C. F. (1969). *Tetrahedron Letters*, 4941.

Hogeveen, H., and Roobeek, C. F. (1970). *Rec. Trav. Chim.* 89, 1121.

Hogeveen, H., and Roobeek, C. F. (1971a). Unpublished results.

Hogeveen, H., and Roobeek, C. F. (1971b). *Tetrahedron Letters*, 3343.

Hogeveen, H., and Roobeek, C. F. (1972). *Rec. Trav. Chim.* 91, 137.

Hogeveen, H., Bickel, A. F., Hilbers, C. W., Mackor, E. L., and Maclean, C. (1967). *Rec. Trav. Chim.* 86, 687.

Hogeveen, H., Lukas, J., and Roobeek, C. F. (1969). *Chem. Commun.* 920.

Hogeveen, H., Baardman, F., and Roobeek, C. F. (1970). *Rec. Trav. Chim.* 89, 227.

Jansen, H. B., and Ros, P. (1969). *Chem. Phys. Letters* 3, 140.

Kaplan, F., Cross, P., and Prinstein, R. (1970). *J. Am. Chem. Soc.* 92, 1445.

Lukas, J., and Kramer, P. (1971) to be published.

Modena, G., and Tonellato, V. (1971). *Adv. Phys. Org. Chem.* 9, 185.

Ohashi, K., Suzuki, S., and Ito, H. (1952). *J. Chem. Soc. Japan, Ind. Chem. Sect.* 55, 120.

Olah, G. A., and Kuhn, S. J. (1964) in "Friedel-Crafts and Related Reactions" (G. A. Olah, ed.), Vol. III, Part 2, p. 1153, Interscience Publishers, New York.

Olah, G. A., and Schlosberg, R. H. (1968). *J. Am. Chem. Soc.* 90, 2726.

Olah, G. A., Baker, E. B., Evans, J. C., Tolgyesi, W. S., McIntyre, J. S., and Bastien, I. J. (1964). *J. Am. Chem. Soc.* 86, 1360.

Olah, G. A., Klopman, G., and Schlosberg, R. H. (1969). *J. Am. Chem. Soc.* 81, 3261.

Olah, G. A., White, A. M., DeMember, J. R., Commeyras, A., and Lui, C. Y. (1970). *J. Am. Chem. Soc.* 92, 4627.

Peters, J. A., and van Bekkum, H. (1971). *Rec. Trav. Chim.* 90, 65.

Pincock, R. E., Grigat, E., and Bartlett, P. D. (1959). *J. Am. Chem. Soc.* 81, 6332.

Raber, D. J., Bingham, R. C., Harris, J. M., Fry, J. L., and Schleyer, P. v. R. (1970). *J. Am. Chem. Soc.* 92, 5977.

Richey, H. G., and Richey, J. M. (1970) in "Carbonium Ions" (G. A. Olah and P. v. R. Schleyer, eds.), Vol. 2, p. 899, Interscience Publishers, New York

Saunders, M., Hagen, E. L., and Rosenfeld, J. (1968). *J. Am. Chem. Soc.* 90, 6882.

Schleyer, P. v. R., Lam, L. K. M., Raber, D. J., Fry, J. L., McKervey, M. A., Alford, J. R., Cuddy, B. D., Keizer, V. G., Geluk, H. W., and Schlatmann, J. L. M. A. (1970). *J. Am. Chem. Soc.* 92, 5246.

Skell, P. S., and Starer, I. (1959). *J. Am. Chem. Soc.* 81, 4117.

Stork, G., and Bersohn, M. (1960). *J. Am. Chem. Soc.* 82, 1261.

Sugita, N., Yasutomi, T., and Takezaki, Y. (1970). *Bull. Japan Petrol. Inst.* 12, 66.

Vegter, G. C. (1970). *Chem. Ind.* 1461.

CHEMICALLY INDUCED DYNAMIC NUCLEAR SPIN POLARIZATION AND ITS APPLICATIONS

D. BETHELL AND M. R. BRINKMAN

The Robert Robinson Laboratories, University of Liverpool, Liverpool, England

I. Introduction	. . . .	53
A. Scope	. . . .	53
B. The N.M.R. Experiment	. . . .	54
C. Dynamic Nuclear Polarization	. . . .	55
II. The Radical Pair Theory of CIDNP	. . . . . .	56
A. Basic Concepts	. . . .	56
B. The Relative Energies of Ground-State T and S Manifolds of a Radical Pair		61
C. The Dynamic Behaviour of Radical Pairs	. . . . .	63
D. Mechanisms of T–S Mixing	. . . .	65
E. Quantitative Aspects of T_0–S Mixing	. . . . . .	68
III. Qualitative Analysis of CIDNP Spectra	. . . . .	73
IV. CIDNP at Low Magnetic Fields	. . . . . .	76
V. Applications of CIDNP	. . . .	78
A. Scope and Limitations	. . . .	78
B. Reactions of Diacyl Peroxides and Related Compounds	. . .	82
C. Reactions of Azo-, Diazo-, and Related Compounds	. . .	95
D. The Photolysis of Aldehydes and Ketones	. . . . .	104
E. Reactions Involving Organometallic Compounds	. . . .	110
F. Ylid and Related Molecular Rearrangements	. . . .	115
VI. Chemically Induced Dynamic Electron Spin Polarization (CIDEP)	. .	120
VII. Prospects	. . . .	121
References	. . . .	122

I. INTRODUCTION

A. *Scope*

THE first reports of the observation of transient emission and enhanced absorption signals in the ^{1}H-n.m.r. spectra of solutions in which radical reactions were taking place appeared in 1967. The importance of the phenomenon, named Chemically Induced Dynamic Nuclear Spin Polarization (CIDNP), in radical chemistry was quickly recognized. Since that time, an explosive growth in the number of publications on the subject has occurred and CIDNP has been detected in ^{2}H, ^{13}C, ^{15}N, ^{19}F and ^{31}P as well as ^{1}H-n.m.r. spectra. Nevertheless, the number of groups engaged in research in this area is comparatively small. This may be a consequence of the apparent complexity of the subject. It is the purpose of this review to describe in a qualitative way the origin of CIDNP and to survey the published applications of the phenomenon in studies of organic reactions.

The origin of CIDNP lies in the microscopic behaviour of radical pairs. Our discussion of this will follow fairly closely the "model" approach associated with the names of Closs, Kaptein, Oosterhoff, and Adrian, rather than the more formal kinetic treatments of Fischer (1970a) and Buchachenko *et al.* (1970b).

Analysis of CIDNP spectra can yield such diverse information as:

(i)   the occurrence of radical pathways in chemical reactions, even if they represent only a minor proportion of the total reaction;

(ii) the nature of the pathway from radicals to final products;

(iii) the electronic multiplicity of reactants and reactive intermediates such as photo-excited states or carbenes;

(iv) the absolute signs and in some cases magnitudes of electron–nuclear (hyperfine) coupling constants and nuclear–nuclear (spin–spin) coupling constants;

(v)   relative magnitudes of radical g-factors;

(vi) information on spin-lattice relaxation in radicals and reaction products;

(vii) energy differences between singlet and triplet states of radical pairs.

Potential applications of CIDNP thus span the whole breadth of chemistry from organic reactions to the physical chemistry of the liquid state. A progress report seems timely. Other, rather more limited reviews have appeared (Closs, 1971b; Fischer, 1971a; Buchachenko and Zhidomirov, 1971; Iwamura, 1971; Ward, 1972; Lawler, 1972).

B. *The N.M.R. Experiment*

As a basis for subsequent discussion, we begin with a brief outline of relevant aspects of the normal n.m.r. experiment in which an assembly of nuclei with half-integral spins is observed (for a fuller treatment of the basic principles of magnetic resonance, see e.g. Carrington and McLachlan, 1967).

When nuclei with spin $\frac{1}{2}$ are placed in a magnetic field, they distribute themselves between two Zeeman energy states. At thermal equilibrium the number (N) of nuclei in the upper (α) and lower (β) states are related by the Boltzmann equation (1) where $\Delta E = E_\alpha - E_\beta$ is the energy difference between the states. In a magnetic field (H_0), $E_\alpha = \frac{1}{2}\gamma\hbar H_0$ and

$$N_\alpha/N_\beta = e^{-\Delta E/kT} \tag{1}$$

$E_\beta = -\frac{1}{2}\gamma\hbar H_0$, γ being the magnetogyric ratio of the nuclei. Thus $\Delta E = \gamma\hbar H_0$. Transitions between the two states are brought about by application of the resonance frequency, $\nu = \Delta E/\hbar = \gamma H_0$, and absorption of energy can occur. This gives rise to the normal n.m.r. spectrum.

Since the total number of nuclei $N = N_\alpha + N_\beta$, it is evident from equation (1) that $(N_\beta - N_\alpha)$, and hence the intensity of the n.m.r. signal, is proportional to N. Typically, the excess population in the β-state is $1:10^5$. Application of the appropriate radiofrequency field induces both upward and downward transitions, but the former predominate. The observed signal indicating energy absorption would quickly be saturated $(N_\beta = N_\alpha)$ but for a non-radiative process, spin-lattice relaxation, by which the Boltzmann distribution can be continuously re-established. Spin-lattice relaxation times, that is the time required for a collection of nuclei to return to the Boltzmann distribution after perturbation, varies according to the type of nucleus (e.g., ^1H, ^{13}C, etc.) and, for a given type, according to its chemical and magnetic environment.

Clearly, if a situation were achieved such that N_α exceeded N_β, the excess energy could be absorbed by the rf field and this would appear as an emission signal in the n.m.r. spectrum. On the other hand, if N_β could be made to exceed N_α by more than the Boltzmann factor, then enhanced absorption would be observed. N.m.r. spectra showing such effects are referred to as polarized spectra because they arise from polarization of nuclear spins. The effects are transient because, once the perturbing influence which gives rise to the non-Boltzmann distribution (and which can be either physical or chemical) ceases, the thermal equilibrium distribution of nuclear spin states is re-established within a few seconds.

C. *Dynamic Nuclear Polarization*

Methods of disturbing the Boltzmann distribution of nuclear spin states were known long before the phenomenon of CIDNP was recognized. All of these involve multiple resonance techniques (e.g. INDOR, the Nuclear Overhauser Effect) and all depend on spin-lattice relaxation processes for the development of polarization. The effect is referred to as dynamic nuclear polarization (DNP) (for a review, see Hausser and Stehlik, 1968). The observed changes in the intensity of lines in the n.m.r. spectrum are small, however, reflecting the small changes induced in the Boltzmann distribution.

Although it is now established that CIDNP has a quite different origin from DNP, the two effects were initially thought to be related. Thus the Overhauser effect, in which saturation of unpaired electron spins leads to polarization of nuclei coupled to the electrons through the hyperfine coupling constant (a_N), can be observed in organic radicals, and CIDNP

3

is observed during reactions involving radical intermediates. However, polarization in the Overhauser case is developed through nuclear relaxation processes, in particular an electron–nuclear cross-relaxation mechanism which requires that when the electron spin changes from β to α, the coupled nuclear spin must change from α to β. It can be shown that, if this were the only relaxation process, the maximum achievable difference in nuclear spin populations $(N_\beta - N_\alpha)$ is increased by an amount determined by the electron- and nuclear-g factors and magnetons (β_E and β_N), namely $(1 + g_E \beta_E / g_N \beta_N) = 659$. Because this factor is too small to explain CIDNP effects and for other reasons described in Section II, electron–nuclear cross-relaxation is no longer regarded as responsible for CIDNP. However, it remains possible that Overhauser-type effects make a small contribution.

II. THE RADICAL PAIR THEORY OF CIDNP

A. *Basic Concepts*

The beginning of CIDNP was the observation (Bargon *et al.*, 1967; Bargon and Fischer, 1967; Ward and Lawler, 1967; Lawler, 1967) of transient emission and enhanced absorption signals during chemical reactions carried out in the magnetic field of a high resolution n.m.r. spectrometer. The frequency parameters of polarized and depolarized spectra for a given chemical species, i.e. chemical shift and nuclear–nuclear spin coupling constants, were identical but line intensities were changed, the most notable line intensity alteration being strong emissions in certain cases. Typical spectra are shown in Fig. 1.

Quite obviously these spectral effects resulted from a non-Boltzmann distribution of nuclear spin populations. The first explanation of such observations (Bargon and Fischer, 1967; Lawler, 1967) consisted of a mechanism involving electron–nuclear cross-relaxation in free radicals. This was an obvious starting point since, as discussed in Section I.C, Overhauser effects were known to produce polarized spectra. However, polarization by such a mechanism could not explain the following features of CIDNP spectra: (i) both emissions and absorptions were observed within a multiplet of a given nucleus or nuclei (Ward and Lawler, 1967; Lawler, 1967; Lepley, 1968, 1969b; Kaptein, 1968); (ii) patterns of polarization were dependent on the type of reaction product (Kaptein, 1968) and multiplicity of the electronic state of the precursor (Closs and Closs, 1969; Closs, 1969; Closs and Trifunac, 1969; Kaptein *et al.*, 1970); (iii) polarization also occurred when reactions were run at zero magnetic field (Ward *et al.*, 1969c; Lehnig and Fischer, 1969); (iv) the magnitude of the enhancements were found in some cases to be greater than the

Overhauser limit (Closs and Closs, 1969; Closs, 1969; Closs and Trifunac, 1969); (v) polarization was observed in radical systems where the radical lifetimes were longer than the nuclear relaxation times in the individual radicals involved (Ward and Lawler, 1967; Lawler, 1967; Fischer, 1969).

All these observations necessitated theorists to discard the Overhauser explanation. The radical pair model was put forward (Closs and Closs,

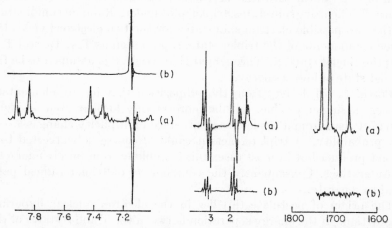

FIG. 1. N.m.r. spectra showing CIDNP. (i) ^{1}H-Spectrum (100 MHz) obtained during (a) and after (b) thermal decomposition of 4-chlorobenzoyl peroxide in hexachloroacetone at 125°, showing net polarization (E) of p-dichlorobenzene and the normal spectrum of the peroxide (Blank and Fischer, 1971b). (ii) ^{1}H-Spectrum (60 MHz) obtained during (a) the reaction of ethyl lithium with ethyl iodide in benzene at 40° showing multiplet polarization (A/E) of the methyl and methylene protons of ethyl iodide. (The other peaks are due to butane and the methylene signal is shown with double the amplitude of the rest of the spectrum.) A reference spectrum of ethyl iodide is shown in (b) (Ward et al., 1969b). (iii) ^{19}F-Spectrum (56·4 MHz) obtained during (a) and after (b) reaction of diphenyl-methylene with benzyl fluoride solvent at 120°, showing combined net and multiplet polarization (A + A/E) of 1,2,2-triphenylethyl fluoride (Bethell et al., 1972a). Scales show chemical shifts (ppm) downfield from tetramethylsilane (i and ii) and (Hz) from $Cl_3C.CF_3$ (iii).

1969; Closs, 1969; Closs and Trifunac, 1969, 1970a; Kaptein and Ooster-hoff, 1969a, b) and this has proved very successful in explaining the experimental findings.

The basic postulates of radical pair theory may be summarized as follows:

(i) Radical pairs must be formed, and these will exist in either electronic singlet or triplet states;

(ii) only in singlet radical pairs can the component radicals react together to give products; triplet radical pairs tend to diffuse apart;

    (iii) radical pairs undergo intersystem crossing (T–S mixing) induced by differences in the radical g-factors and electron–nuclear hyperfine interactions: this process gives rise to nuclear spin selection.

Postulate (i) follows from the fact that when two radicals, produced by whatever means, encounter each other, the interaction of the electron spin of one radical with that on the other radical can give rise to two mutually exclusive spin states, triplet and singlet. Random combination of the two possible electron spin states for the two electrons yields the three components of the triplet state, represented as T_{+1}, T_0, and T_{-1}, and the singlet state, S. Throughout this article, S is assumed to be the singlet state of lowest energy.

Postulate (ii) derives from the proposition that in any elementary process a singlet product will be more readily formed from a singlet reactant than from a triplet reactant, since multiplicity changes are of low probability. A triplet reactant could of course be converted to a triplet product but in most cases this is unlikely from purely energetic considerations. Consequently the components of triplet radical pairs tend to separate.

The origin of postulate (iii) lies in the electron–nuclear hyperfine interaction. If the energy separation between the T and S states of the radical pair is of the same order of magnitude as a_N, then the hyperfine interaction can represent a driving force for T–S mixing and this depends on the nuclear spin state. Only a relatively small preference for one spin-state compared with the other is necessary in the T–S mixing process in order to overcome the Boltzmann polarization (1 in 10^5). The effect is to make n.m.r. spectroscopy a much more sensitive technique in systems displaying CIDNP than in systems where only Boltzmann distributions of nuclear spin states obtain. More detailed consideration of postulate (iii) is deferred until Section II,D.

Accepting all three of the basic postulates, the origin of CIDNP can now be described qualitatively. Figure 2 illustrates a variety of different radical reaction pathways known to produce polarized spectra.

Correlated or geminate radical pairs are produced in unimolecular decomposition processes (e.g. peroxide decomposition) or bimolecular reactions of reactive precursors (e.g., carbene abstraction reactions). Radical pairs formed by the random encounter of freely diffusing radicals are referred to as uncorrelated or encounter (F) pairs. Once formed, the radical pairs can either collapse, to give combination or disproportionation products, or diffuse apart into free radicals (doublet states). The free radicals escaping may then either form new radical pairs with other radicals or react with some diamagnetic scavenger molecule SX. A precursor, M, with a given electronic spin state multiplicity, will give initially a radical pair of the same multiplicity, singlet (S) or triplet (T). Random free radical encounters forming radical pairs could yield either S or T states. Products from S and T precursors can form either cage collapse product (c) or escape products (e).

In order to explain qualitatively how CIDNP arises, the simple case of a radical pair in which only one component contains a nucleus (spin $\frac{1}{2}$) coupled to an electron (spin $\frac{1}{2}$) through a hyperfine coupling constant a_N will be considered.

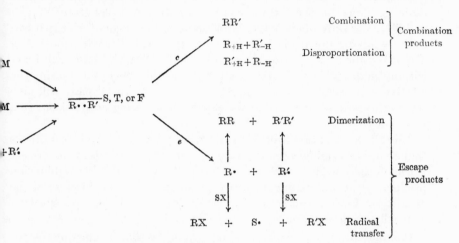

FIG. 2. Formation and reactions of radical pairs. M = precursor (multiplicity, S or T); SX = solvent molecule or radical scavenger.

If the radical pair is formed from a triplet precursor ^TM, then the radical pair will initially be in the triplet manifold. The cross-over between the T and S manifolds of the radical pair is denoted by T–S mixing. The T–S mixing process has the properties that one particular nuclear spin configuration α or β will be selected and that the same configuration will be selected whether going from T to S or S to T. In addition, only the S manifold of a radical pair can give c-products. Therefore, after entering through the T manifold initially, the time development will yield some S radical pairs. If the T–S mixing preferentially selects the β spin state, S radical pairs will be enriched in the β spin state and, by difference, T radical pairs in the α spin state.

Since only singlet radical pairs can give c-products the c-products are enriched in β spin, giving enhanced absorption (A), and e-products are enriched in α spin, leading to emission (E). The immediate result is that CIDNP spectra can distinguish c- and e-products from a precursor of specified multiplicity.

If the radical pair had been created initially through the S manifold, T–S mixing would again select the β spin states. Thus the time development would yield T radical pairs and e-products enriched in β spin states whereas S radical pairs and c-products would be enriched in the α spin state. CIDNP spectra can therefore distinguish between T and S precursors.

Consider now an encounter (F) radical pair formed from two free radicals. Since there are three components to the triplet state, T_{+1}, T_0 and T_{-1}, and only one singlet component, S, the encounter of two free radicals having uncorrelated spins leads to a statistical distribution of T and S radical pairs. However, some of the S radical pairs will react without undergoing T–S mixing, and this has the effect of increasing the relative number of T radical pairs. Consequently the F-pairs will give the same type of polarization as the T-pairs, but the degree of polarization will be less.

There is an additional problem that must be considered for radicals that escape an original geminate radical pair. Such radicals are polarized. If they form new (encounter) radical pairs, they could yield combination products after undergoing T–S mixing again. However, because of the rapidity of nuclear relaxation in free radicals, the F-type polarization generally predominates.

It should be evident from the above examples that, neglecting relaxation, if the polarization of all products in the system is considered, there is no overall net polarization: only individual products show net polarization.

Provided that polarization arises predominantly as a result of the difference in the g-factors of the components of the radical pair, similar considerations apply to systems where several coupled nuclei are polarized. Net polarization, A or E, results. This is indicated in Fig. 3b for the AB system of protons in a hypothetical product $R_2CH_A . CH_BR_2'$ formed as a c-product from the singlet radical pair $\overline{R_2CH_A \cdot \cdot CH_BR_2'}^S$. However, if $\Delta g \approx 0$ for the pair, hyperfine interactions dominate during T–S mixing. As a result, depending on the absolute signs of a_{H_A}, a_{H_B}, and J_{AB}, the $\alpha\alpha$ and $\beta\beta$ spin states are overpopulated compared with the $\alpha\beta$ and $\beta\alpha$ states, or *vice versa*. The observed spectrum of the multiplet then shows both absorption and emission lines with an integral over the multiplet of zero (Fig. 3c, d). This is referred to as multiplet polarization or simply the multiplet effect. This sort of polarization is designated A/E or E/A according as the signal at lowest field shows absorption or emission. Often, net and multiplet polarization occur together in the same multiplet, the superimposed effects resulting in an unsymmetrical combination of absorption and emission signals having a non-zero integral. Combination of, for example, net absorption with an A/E multiplet effect is designated A + A/E (See Fig. 1c).

With this qualitative description for guidance, the underlying physical processes can now be examined in greater detail.

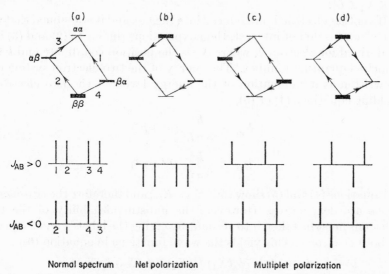

FIG. 3. Schematic representation of energy levels, populations and resultant patterns of polarization in the n.m.r. spectrum of an AB spin system. (Relative population of the energy levels is indicated by the thickness of the bars.)

B. *The Relative Energies of Ground State T and S Manifolds of a Radical Pair*

The approach to this problem (McGlynn *et al.*, 1969) will follow the general quantum-mechanical treatment of the problem of two particles in a box. For convenience, the two particles are taken to be electrons. The two particles of interest in CIDNP are two free radicals, but the conclusion reached concerning the relative energies of T and S states should be the same. The size of the box will represent the separation of the electrons in the radical pair. For simplicity, first consider the energy levels of a single electron confined to a one-dimensional box of length L. These are given by equation (2),

$$E_n = \frac{h^2 n^2}{8mL^2} ; \qquad n = 1, 2, 3 \ldots \tag{2}$$

where E_n is the energy of the level with principal quantum number n, h is Planck's constant, and m the relativistic mass of the electron. The appropriate wavefunction is given by equation (3),

$$\phi_n(X_1) = \left(\frac{2}{L}\right)^2 \sin\left(\frac{h\pi X_1}{L}\right) \tag{3}$$

where X_1 represents the position coordinate of the electron in the box $(0 \leqslant X_1 \leqslant L)$.

If another electron is introduced into the box and it is assumed that the two electrons do not interact, then expressions similar to (2) and (3) can be derived for electron 2, where X_2 is the position coordinate and k the principal quantum number. The energy of the two-electron system can be written as a summation of the energy levels of the two electrons, yielding equations (4) or (5),

$$E_{nk} = \frac{h^2}{8mL^2}(n^2 + k^2) \tag{4}$$

$$E_{kn} = \frac{h^2}{8mL^2}(k^2 + n^2) \tag{5}$$

Equations (4) and (5) show that $E_{nk} = E_{kn}$, and therefore the two energy levels are degenerate. However, the indistinguishability of the two electrons must be taken into account by giving the labels 1 and 2 equally to both electrons. This yields the wave functions in equation (6),

$$\psi_{\pm} = 2^{-1/2}[\phi_n(X_1)\,\phi_k(X_2) \pm \phi_n(X_2)\,\phi_k(X_1)] \tag{6}$$

with $2^{-1/2}$ the normalization factor. The label changing can be done in two ways with respect to the electron coordinate system (i) $\psi_+ \rightarrow \psi_+$ or (ii) $\psi_- \rightarrow -\psi_-$ representing a symmetric and antisymmetric process, respectively.

In the absence of any interacting forces between the two electrons it still remains that $E_+ = E_- = E_{nk} = E_{kn}$. To remove the degeneracy of the energy levels one needs to include an interaction between the two electrons, an appropriate choice being the Coulomb repulsive force with an energy given by $e^2/X_{12} = e^2/|X_1 - X_2|$ (e is the electronic charge and X_1 and X_2 are the coordinate vectors in the three-dimensional case). Two energy terms have thus been added, namely the Coulomb term, K_{nk}, and the electron exchange energy J_{nk}. If both K_{nk} and J_{nk} are taken to be positive, it can be seen that K_{nk} will raise the energy of the system and J_{nk} will either raise or lower the energy of the system [equations (7) and (8)].

$$E_{\pm} = E_{nk} + K_{nk} \pm J_{nk} \tag{7}$$

$$E_+ - E_- = 2J_{nk} \tag{8}$$

Since two electrons with symmetric space wavefunctions and antisymmetric space wavefunctions represent singlet and triplet states respectively, then obviously the triplet state (E_-) is of lower energy than the singlet state (E_+) by an amount $2J_{nk}$. Had an attractive force between the two electrons been assumed (e.g. a gravitational force), the triplet state would have been found to be of higher energy.

It should be noted that the above conclusions have been reached on strictly electrostatic grounds; a spin property has not been invoked for the two electrons. From the variation of $\psi_\pm^2$ along the box it can be shown that the singlet state is of higher energy than the triplet because the two electrons are more crowded together for ψ_+^2 (S-state) than for ψ_-^2 (T-state). Thus there is less interelectronic repulsion in the T-state. The quantity $2J_{nk}$ is a measure of the effect of electron correlation which reduces the repulsive force between the two electron (Fermi correlation energy).

Introduction of the half-integral spin of the electrons (values $\hbar/2$ and $-\hbar/2$) alters the above discussion only in that a spin coordinate must now be added to the wavefunctions which would then have both space and spin components. This creates four vectors (three space and one spin component). Application of the Pauli exclusion principle, which states that all wavefunctions must be antisymmetric in space and spin coordinates for all pairs of electrons, again results in the T-state being of lower energy [equations (9) and (10)].

$$E(S_i) > E(T_i) \tag{9}$$
$$\Delta E_{ST} = E(S_i) - E(T_i) = 2J_{nk} \tag{10}^1$$

The principal concern in radical pair theory is not which state is of lower energy but rather the energy difference ΔE_{ST} and its variation as the length of the box or the inter-radical distance in the radical pair changes. Further, ΔE_{ST} is time-dependent because the inter-radical distance fluctuates with time. As the separation gets larger, J_{ee} decreases and consequently so does ΔE_{ST}. At a low enough value of J_{ee}, crossover between S and T manifolds of the radical pair would be energetically favourable if there were a driving force equal to ΔE_{ST}. The hyperfine field a_N can provide this driving force if equation (11) is satisfied.

$$\Delta E_{ST} = E(S_i) - E(T_i) = J_{ee} = a_N \tag{11}$$

C. The Dynamic Behaviour of Radical Pairs

A simple model (Fig. 4) will assist in visualizing the time-dependent variation of the separation and hence the interaction of the components of a radical pair.

---

1 Henceforth the quantity $2J_{nk}$ will be referred to as simply J_{ee}.

In Fig. 4, nucleus H_i is coupled to the electron on R and nucleus H_j is coupled to the electron on R′ by the hyperfine terms a_i and a_j, respectively. The inter-radical separation is denoted by X, and J_{ee} is the electron exchange integral. The two radicals move relative to each other and, as shown above, this changes the values of J_{ee}.

Consider a geminate radical pair, its component radicals in close proximity and therefore having correlated electron spins reflecting the electronic spin state of the precursor. Three different situations can be envisaged.

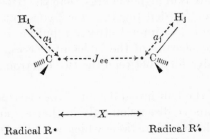

Radical R·   Radical R′·

FIG. 4. Schematic representation of a radical pair showing the important parameters.

Case (1)

Immediately after formation of the geminate pair (time, t_1) the component radicals are close together; X_1 is small and J_{ee} and therefore ΔE_{ST}, large. Thus $J_{ee} \gg a_i \approx a_j$.

Case (2)

At a later time t_2, the two radicals will have diffused apart slightly or $X_2 > X_1$ which implies that $J_{ee}(t_2) < J_{ee}(t_1)$. Since $\Delta E_{ST}(t_2) < \Delta E_{ST}(t_1)$, t_2 can be chosen so that the value of $J_{ee} \simeq a_i \simeq a_j$.

Case (3)

At a later time t_3, the two radicals could still correspond to Case (2), or have returned to the situation in Case (1), or they could have diffused so far apart that $J_{ee} = 0$. In the last situation, the electron spins are no longer correlated and the radicals are free (doublet states).

At time t_2 [Case (2)] therefore, the hyperfine energy is approximately equal to the energy difference between the S and T states and can provide the driving force for T–S mixing. Now the hyperfine constants a_i and a_j are a function of both nuclear and electronic spin states and thus one particular nuclear spin state for H_i and H_j will induce the T–S mixing more readily than the other. Thus nuclear spin selection occurs during the transition between S and T manifolds. However, this would yield no net polarization in either c- or e-products if it were not that *only* singlet radical pairs yield c-products.

D. *Mechanisms of T–S Mixing*

1. *A general mechanism*

Inter-system crossing involves a radiationless transition between electronic singlet and triplet manifolds. So far, pure electronic T and S states have been assumed, i.e. it has been assumed that the spin angular momentum of a state does not vary with time. However, a transition between two such pure spin states requires a change in spin angular momentum. Now momentum is a physical quantity that is always conserved, and since the operators representing the applied magnetic and electric fields are not a function of spin coordinates, the applied magnetic field cannot absorb such a change in spin angular momentum.

Spin-orbit coupling, however, provides a mechanism by which momentum can be conserved. Put simply, the angular momentum is allowed to be exchanged between spin and orbital degrees of freedom while still conserving total (i.e., spin plus orbital) angular momentum. Unfortunately, the pleasantly simple description of pure spin states is lost in the process; there are still real electronic spin states, but each must be described as containing a small amount of the other. Spin–orbit coupling does provide a mechanism for T–S mixing, but for CIDNP it is not as important as that provided by the hyperfine field.

2. T_0 –S *mixing in radical pairs*

So far the roles of the three components of the triplet manifold $(T_0, T_{\pm 1})$ have not been considered. In Fig. 5 the relationship between the energies of the three components of the T state and the S state is shown.

As can be seen from Fig. 5, both the T_0 and T_{-1} states can mix with the S state but at different inter-radical distances X. At the point $J_{ee} \approx \omega_s$, X is small. The radical pair will not spend much time in this region; either it will collapse to give a c-product or the two radicals will diffuse apart to the region where $J_{ee} \approx a$ and T_0–S mixing becomes possible. T_{-1}–S mixing is thus rather uncommon except where the radical separation is restricted by complexing or in a diradical (see Section IV).

Only T_0–S mixing will be considered now since it is responsible for most CIDNP spectra, certainly at zero and at high magnetic fields.

A pictorial representation of the T_0–S mixing process follows from Fig. 6. Just as in normal n.m.r. or e.s.r. spectroscopy, precession can be represented by a vector model. When placed in an external magnetic field the two unpaired electrons of the radical pair 1 and 2 will precess

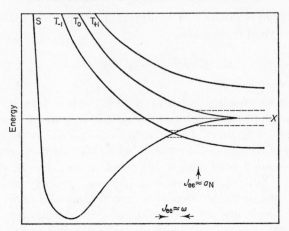

FIG. 5. Energies of S, T_0 and $T_{\pm 1}$ states of a radical pair as a function of radical separation. The dotted lines indicate mixing of states.

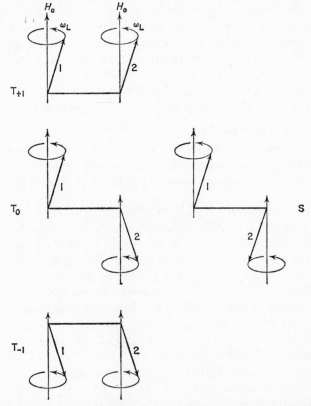

FIG. 6. Precession of electron spins in S, T_0 and $T_{\pm 1}$ states of a radical pair.

about the magnetic field axis at definable phase angles. The phase angles
are different for the four possible electronic states of the radical pair,
T_{+1}, T_0, T_{-1} and S. The frequency with which the two electrons
precess is the familiar Larmor precession frequency (LPF). There is
no crossing over between any of the four states if the LPFs of electrons 1
and 2 remain equal. However, it can be seen from Fig. 6 that if electron
2 in the T_0 state were to precess a little faster than electron 1, then
after a while it would have changed phase by 180° and the radical pair
would be in the S state. Therefore T_0–S mixing is dependent on the
electrons having different LPFs. For a radical in a high magnetic field
the LPF (in angular frequency units) is dependent upon the g-factor
and the hyperfine interactions between the electron and nuclei as
shown in equation (12).

$$\omega_L = g\beta H_0 + \sum^i a_i m_{Ii} \tag{12}$$

Thus both the g-factor and the a_i-values affect T_0–S mixing. In general,
the two radicals comprising the radical pair will have different g-values
and therefore different Zeeman energies [the first term in equation (12)].
Likewise the a_i-values and hyperfine energies [the second term in equation (12)] will differ. The difference between the LPFs for two radicals is
then given by equation (13), where g_1 and g_2 are the electronic g-factors

$$\omega_{L1} - \omega_{L2} = (g_1 - g_2)\beta H_0 + a_1 m_1 - a_2 m_2 \tag{13}$$

of radicals 1 and 2, a_1 and a_2 the hyperfine constants, and m_1 and m_2
nuclear spin quantum numbers. It should be noted that the first term
in equation (13) represents a magnetic field (H_0) dependence and the
remaining terms a nuclear spin state dependence, from which nuclear
spin selection is derived.

It can also be seen from Fig. 6 that if the T_{+1} or T_{-1} states mixed with
S, this would involve concomitant electron and nuclear spin flipping in
order that the total spin angular momentum be conserved, and this
would ultimately produce the same polarization in c- and e-products. This
point will be discussed further in Section IV.

3. The effect of the life-times of radical pairs on the magnitude of polariza-
tion

The lifetime of the radical pair is important since successive encounters
of radicals [going between case (1) and case (2)] will give more T_0–S
mixing and therefore greater nuclear spin polarization.

The problem of estimating the number of encounters that a given pair
of radicals will have is not trivial. The most successful approach
(Adrian, 1970, 1971; Kaptein, 1971b, 1972a, b; Kaptein and den

Hollander, 1972) using basically the Noyes diffusion model (Noyes, 1954, 1955, 1956, 1961), treats the radicals as if they go on a "random walk". It must be remembered, however, that the two radicals must maintain the correlation of their electron spins ($J_{ee} \neq 0$) or the distinction between T and S manifolds for a given radical pair would be lost.

Furthermore, even though the time dependence of the mixing process has been placed in J_{ee}, J_{ee} must approximately equal a_i during successive encounters in order that the time development of nuclear spin states in

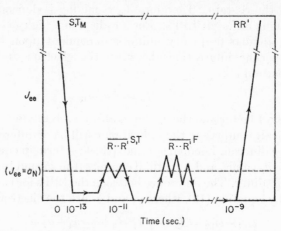

FIG. 7. Variation of the electron exchange energy for radical pairs during a "random walk".

the products via T_0–S mixing be adiabatic. In Fig. 7 is a representation of the random walk process in which the value of J_{ee} drops immediately to a value approximately equal to a_i and remains so during successive encounters. A more quantitative treatment of CIDNP will be given in the next section. The approach is essentially that of Kaptein (1971b, 1972a, b; Kaptein and den Hollander, 1972).

E. Quantitative Aspects of T_0–S Mixing

1. The spin Hamiltonian and T_0–S mixing

A basic problem in quantum mechanics is to relate the probability of an ensemble of particles being in one particular state at a particular time to the probability of their being in another state at some time later. The ensemble in this case is the population distribution of nuclear spin states. The time-dependent Schrödinger equation (14) allows such a calculation to be carried out. In equation (14) $\psi(S,t)$ denotes the total

$$i\frac{\partial \psi(S,t)}{\partial t} = \mathscr{H}(S,t)\psi(S,t) \tag{14}$$

wavefunction for electron and nuclear spin coordinates, and $\mathscr{H}(S,t)$ the time-dependent Hamiltonian.

It is necessary to substitute a phenomenological equation for $\mathscr{H}(S,t)$, which in this case involves only spin variables.

The spin Hamiltonian is thus generated. In particular it can be used to examine the T_0–S mixing of electron spin states and its relationship to the distributions of populations of nuclear spin states. The total spin Hamiltonian is given in equation (15) which contains both electron and nuclear terms.

$$H = H_e + H_{LS} + H_{HL} + H_{HS} + H_{SS} + H_{SI} + H_{HI} \qquad (15)$$

The first term is the electronic part of the Hamiltonian and this can be expanded as shown in equation (16).

$$H_e = H_e^A + H_e^B + H_e^{AB} \qquad (16)$$

Here H_e^A and H_e^B describe radicals A and B of the radical pair and H_e^{AB} the interaction of their electrons. The other terms in equation (15) are H_{LS}, the spin orbit coupling term, H_{HS} and H_{HI}, representing the interaction of the externally applied magnetic field with the electron spin and nuclear spin, respectively; H_{SS} is the electron spin–spin interaction and H_{SI} the electron–nuclear hyperfine interaction.

As has been shown (Kaptein, 1971b, 1972a) by application of perturbation theory (Itoh $et\ al.$, 1969), the spin Hamiltonian in equation (17) can be obtained for S and T radical pairs.

$$H_{RP} = H_{ex} + H_{ZS} + H_{hf} + H_D + H_{ZI} \qquad (17)$$

The terms will be considered separately.

The electron–electron exchange term, H_{ex}

In equation (16) it is necessary to consider only H_e^{AB}. As has been discussed, the energy difference between T and S states is equal to J_{ee}. With a minimal overlap integral due to a relatively large inter-radical separation, H_{ex} can be given by the Dirac exchange operator [equation (18)],

$$H_{ex} = -J_{ee}(\tfrac{1}{2} + 2S_1 \cdot S_2) \qquad (18)$$

where S_1 and S_2 are the two electron spin operators. For two electrons in orbitals ψ_A and ψ_B, the exchange integral J_{ee} is given by equation (19).

$$J_{ee} = \langle \psi_A(1)\psi_B(2) | H_e^{AB} | \psi_B(1)\psi_A(2) \rangle \qquad (19)$$

J_{ee} will fluctuate because of diffusion and tumbling of the radicals in the radical pair. Several workers (Herring and Flicker, 1964; Murrell and Teixeira-Dias, 1970; Hirota and Weissman, 1967; Ferruti $et\ al.$, 1970) have studied the fluctuation of J_{ee} with distance between the electrons and have concluded that on the average J_{ee} will become approximately equal to the hyperfine constant a_N at about a 5–10 Å separation. Such a separation would be possible only after a few diffusive displacements. The time dependence of J_{ee} can be approximated by a step function, however.

The electron–nuclear hyperfine term (H_{hf})

The interaction of electron 1 on radical A with nucleus j (spin operator I_j) and of electron 2 on radical B with nucleus k (spin operator I_k) is given by equation (20)

$$H_{SI} = S_1 . \sum_j^A a_j I_j + S_2 . \sum_k^B a_k I_k \tag{20}$$

The summations are over all nuclei in radicals A and B for which a_j and a_k are finite. The a-values are tensor quantities but are treated as isotropic here.

The residual terms in H_{RP}

The electron–electron dipolar term, H_D, equals $S_1 . D . S_2$. The tensor D is completely anisotropic and only mixes T-states with one another. It is therefore dropped. The nuclear Zeeman term, $H_{ZI} = \sum_i g_N \beta_N I_{Zi} H_0$ can be ignored, as its inclusion would change only the zero point energy without affecting the T_0–S mixing.

Therefore, the effective isotropic spin Hamiltonian for the radical pair H_{RP} is given by equation (21).

$$H_{RP} = H_{ex} + H_{SI} = H^0 + H' \tag{21}$$

The terms in H_{RP} are expressed in equations (22) and (23)

$$H^0 = \tfrac{1}{2}(g_A + g_B)\beta_e \hbar^{-1} H_0(S_{1z} + S_{2z}) - J_{ee}(\tfrac{1}{2} + 2S_1 . S_2)$$
$$+ \tfrac{1}{2}(S_1 + S_2)\left\{ \sum_j^A a_j I_j + \sum_k^B a_k I_k \right\} \tag{22}$$

$$H' = \tfrac{1}{2}(g_A - g_B)\beta_e \hbar^{-1} H_0(S_{1z} - S_{2z})$$
$$+ \tfrac{1}{2}(S_1 - S_2)\left\{ \sum_j^A a_j I_j - \sum_k^B a_k I_k \right\} \tag{23}$$

Only the off-diagonal terms in H_{RP} or only H' need be considered since these are the terms that mix S and T wavefunctions.

2. The time-dependent T_0–S mixing process

The representation in equation (24) will be used for S and T basis functions.

$$S = 2^{-1/2}(\alpha_1 \beta_2 - \beta_1 \alpha_2); \qquad T_0 = 2^{-1/2}(\alpha_1 \beta_2 + \beta_1 \alpha_2) \tag{24}$$

The nuclear spin product functions will be represented by χ_n, where n represents a given nuclear spin configuration characterized by nuclear spin quantum number m_j and m_k for nuclei j and k, respectively.

If $\psi(t)$ represents the wavefunction of the radical pair at time t, then $\psi(t)$ can be assumed to be a function of both the S and T wavefunctions and the nuclear spin states at time t as given in equation (25).

$$\psi(t) = [C_{Sn}(t)\,S + C_{Tn}(t)\,T_0]\,\chi_n \tag{25}$$

where C_{Sn} and C_{Tn} are time-dependent and nuclear spin state-dependent coefficients for S and T states of the radical pair at time t.

Application of the time-dependent Schrödinger equation gives equation (26). This yields two coupled equations which, solved for

$$i\frac{\partial(\psi)}{\partial t} = H_{RP}\,\psi \tag{26}$$

$t = 0$ or $C_S(0)$ and $C_T(0)$, gives equation (27).

$$C_{Sn}(t) = C_S(0)\left(\cos\omega t - \frac{iJ_{ee}}{\omega}\sin\omega t\right) - C_T(0)\,b_n\sin\omega t \tag{27}$$

where $\omega = (J_{ee}^2 + b_n^2)^{1/2}$, the subscript n represents the nuclear spin state, and b_n is given by equation (28).

$$b_n = \langle S\chi_n|H'|T_0\chi_n\rangle \tag{28}$$

Replacing H' by equation (23), equation (29) is obtained.

$$b_n = \frac{1}{2}\left[\Delta g\beta_e\hbar^{-1}H_0 + \sum_j^A a_j m_j - \sum_k^B a_k m_k\right] \tag{29}$$

$$[\Delta g = g_A - g_B]$$

The square of a wavefunction is related to the probability distribution, and therefore the probability of finding the radical pair in the S state is given by equation (30).

$$|C_{Sn}(t)|^2 = |C_S(0)|^2 + \{|C_T(0)|^2 - |C_S(0)|^2\}(b_n/\omega)^2\sin^2\omega t \tag{30}$$

Now, if the precursor was a singlet, it is reasonable to suppose that at $t = 0$, $C_S(0) = 1$ and $C_T(0) = 0$, giving equation (31).

$$|C_{Sn}(t)|^2 = 1 - (b_n/\omega)^2\sin^2\omega t \tag{31}$$

If the precursor were a triplet, then at $t = 0$, $C_S(0) = 0$ and $C_T(0) = 1$, giving equation (32).

$$|C_{Sn}(t)|^2 = (b_n/\omega)^2\sin^2\omega t \tag{32}$$

It is easy to see therefore why nuclear population distributions are opposite for T and S radical pairs. Kaptein has also shown that the assumption that the functions χ_n are eigenfunctions of the precursor is justified. It has also been assumed that the spins of the radical pair remain correlated during diffusive encounter of the geminate pairs and that only $\mathbf{H}'$ alters their phase relationships. This is probably justified because spin-lattice relaxation is slower than recombination of correlated radical pairs. Since recombination occurs only from S radical pairs, the probability of recombination can be taken to be proportional to $|C_{Sn}(t)|^2$.

As mentioned earlier, radicals that do not react can go on a random walk and may meet later at time $t_0 + dt$. Noyes (1954, 1955, 1956, 1961) has shown that the probability of such a re-encounter is $f(t)\,dt$, where $f(t)$ is given by equation (33) and C_{mp} is a complex constant determined

$$f(t) = C_{mp}\,t^{-3/2} \tag{33}$$

by the dynamics of relative motion of the radicals and by the radical dimensions.

If λ is the probability of recombination during a single encounter of a singlet radical pair (close to unity), then the chance of product formation during the second encounter at time t is $\lambda|C_{Sn}(t)|^2 f(t)$. For additional re-encounters, it can be shown that $\lambda_n(t) = \lambda|C_{Sn}(t)|^2$ and therefore recombination during the time interval between t and $t + dt$ has a probability $P_n(t)\,dt$ given by equation (34),

$$P_n(t) = \lambda_n(t)\,f(t) + \int_0^t \lambda_n(t-t_1)[1 - \lambda_n(t_1)]\,f(t-t_1)\,f(t_1)\,dt_1$$

$$+ \int_0^t \lambda_n(t-t_2)\,f(t-t_2)\,dt_2 \int_0^{t_2} [1 - \lambda_n(t_2-t_1)][1 - \lambda_n(t_1)]\,f(t_2-t_1)\,f(t_1)\,dt_1$$

$$+ \dots \tag{34}$$

where t_1, t_2 ... are the times of the first, second, etc. unsuccessful encounters. Approximating, by using only the first one or two terms in equation (34), one can show that the random walk model predicts that the total fraction of radical pairs with a given nuclear state (n) is proportional to $b_n^2/\omega^{3/2}$. Further, opposite polarization is predicted for products from T or F precursors and S precursors.

3. Enhancement factors

The intensity of a non-degenerate n.m.r. transition between the nuclear Zeeman levels n and m is proportional to the difference in population of levels n and m as given by the Boltzmann distribution. This can be expressed by equation (35).

$$I_{nm}^0 \text{ is proportional to } (N_n^0 - N_m^0) = \frac{N}{Z} \cdot \frac{g_N \beta_N H_0}{kT} \tag{35}$$

Here N is the number of molecules present, and the total number of nuclear levels in the given system, $Z = \prod_i (2I_i + 1)$. The superscript zero denotes the state of thermal equilibrium. The enhancement in intensity due to polarization is then given by equation (36), where r is the rate of

$$(N_n - N_m) - (N_n^0 - N_m^0) = rT_{nm}(P_n - P_m) \qquad (36)$$

radical pair formation and T_{nm} a nuclear relaxation time for the n.m.r. transition. Thus the observed enhancement

$$I_{nm} - I_{nm}^0 = \frac{rT_{nm}(P_n - P_m)}{N} \cdot \frac{ZkT}{g_N \beta_N H_0} \qquad (37)$$

Thus, $I_{nm} < 0$ and $I_{nm} > I_{nm}^0$ correspond to emission and enhanced absorption respectively.

III. Qualitative Analysis of CIDNP Spectra

It follows from the discussion in Sections II,E,2 and 3 that if $J_{ee}^2 \gg b_n^2$, then the intensity of an n.m.r. transition between levels n and m, I_{nm}, is proportional to $(b_n^2 - b_m^2)$. From equation (29) for b, it can be shown that for the transition in which nucleus i on component A of the radical pair AB changes its spin from m_i to $m_i - 1$, the intensity I_{mn} is given by equation (38).

$$I_{mn} \text{ is proportional to } \tfrac{1}{2}a_i \left[\varDelta g\beta_e \hbar^{-1} H_0 + \sum_j^A a_j m_j - \sum_k^B a_k m_k \right] \qquad (38)$$

The summations are over all hyperfine interactions in A and B except that with nucleus i.[2]

The first term in equation (39) represents the net effect, E or A, depending on the sign of $a_i \varDelta g$. The intensity is proportional to the magnetic field, H_0, provided $J_{ee}^2 \gg b_n^2$. When this condition does not hold, it turns out that the intensity should decrease with H_0. I_{mn} should thus show a maximum as H_0 increases, since b_n is itself linearly dependent on H_0 (see also Section IV, however).

The remainder of equation (38) describes the multiplet effect, and it can be seen that whether an individual line in the multiplet corresponds to emission or absorption depends on the signs of the hyperfine coupling constants but is independent of H_0. The nature of the hyperfine field is such that the integral over the whole multiplet is zero if $\varDelta g = 0$.

[2] Kaptein's (1971b, 1972a) equivalent expression contains a further term $\tfrac{1}{2}a_i(m_i - \tfrac{1}{2})$ which disappears in first-order spectra (cf. Fischer, 1970; Lehnig and Fischer, 1970; Lawler, 1972).

The polarization shown in Fig. 3c, d for example, can be readily predicted from equation (38) by putting

$$\Delta g = 0; \sum_{j}^{A} a_j m_j = 0; \quad \sum_{k}^{B} a_k m_k = \pm \tfrac{1}{2} a_H^B \quad \text{and} \quad a_i = a_H^A$$

However, the nuclear–nuclear coupling constant also has an influence, since changing its sign changes the assignment of the lines in the spectrum.

Both net and multiplet effects must normally be considered except in two special cases: (i) when $\Delta g = 0$ and only multiplet effects are observed and (ii) when $a_i = 0$ in which case there is no CIDNP to observe. In addition, if there is no coupling between a given nucleus or nuclei and any other nuclei in the product, the n.m.r. spectrum will be a single peak, which of necessity can show only net polarization.

Kaptein (1971a, b) has analysed CIDNP spectra in terms of an expression equivalent to equation (38) and the considerations mentioned above for net and multiplet effects. A summary of his predictions follows:

(i) Polarization arising from S precursors is of opposite sign to that from T and F precursors.

(ii) Escape products (e) and combination products (c) show opposite polarization.

(iii) The multiplet effect is proportional to $a_i^A a_j^A$ or proportional to $-a_i^A a_k^B$ and dependent on the signs of J_{ij} and J_{ik}.

(iv) The net effect is proportional to $\Delta g a_i$.

(v) "Second order" multiplet effects can appear in spectra of magnetically equivalent nuclei, even when the coupling is to a nucleus not present in the radical pair; the phase of the polarization will then depend on the sign of J_{ij} but not on the sign of a_i.

Kaptein (1971a, b, 1972a) has further derived relations for predicting net and multiplet effects on the basis of (i)–(iv) above. The qualitative features of CIDNP spectra can be determined by determining the signs of the functions Γ_{NE} and Γ_{ME} given by equation (39a) and (39b) for net and multiplet effects respectively.

$$\Gamma_{NE} = \mu \epsilon \Delta g a_i \quad \begin{cases} + : A \\ - : E \end{cases} \tag{39a}$$

$$\Gamma_{ME} = \mu \epsilon a_i a_j J_{ij} \sigma_{ij} \quad \begin{cases} + : E/A \\ - : A/E \end{cases} \tag{39b}$$

These hold for nucleus i of radical A, and μ, ϵ and σ_{ij} are labels indicating precursor multiplicity, c or e product, and whether i and j are in the same or different radicals, viz.,

$$\mu \begin{cases} + \text{ for T and F precursors} \\ - \text{ for S precursor} \end{cases}$$

$$\epsilon \begin{cases} + \text{ for } c \text{ products (recombination)} \\ - \text{ for } e \text{ products (escape)} \end{cases}$$

$$\sigma_{ij} \begin{cases} + \text{ when i and j are in the same radical} \\ - \text{ when i and j are in different radicals} \end{cases}$$

It can be seen from (39a and b) that any one of the parameters μ, ϵ, a_i, a_j, J_{ij}, or Δg can be determined if the other parameters are known.

To illustrate the use of (39a and b), two simple examples are given (Kaptein, 1968).

(1) The methoxy group of methyl acetate formed during the thermal decomposition of acetyl peroxide appears as an emission, whereas methyl chloride shows enhanced absorption. Consider the reaction sequence in equation (40).

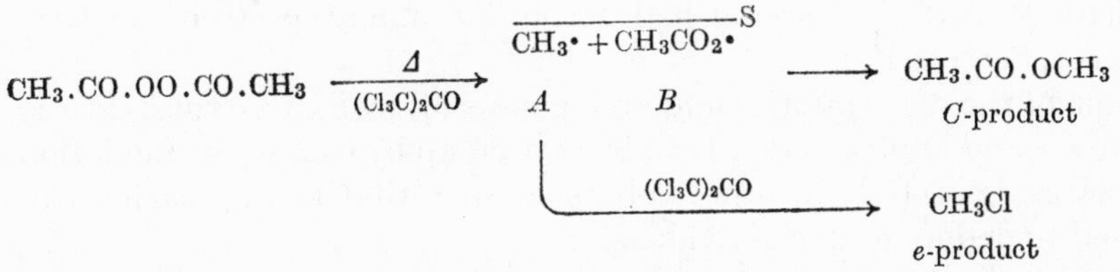

$$CH_3.CO.OO.CO.CH_3 \xrightarrow[\text{(Cl}_3\text{C)}_2\text{CO}]{\Delta} \overset{\text{S}}{\overline{\underset{A \quad\quad B}{CH_3\text{·} + CH_3CO_2\text{·}}}} \longrightarrow \underset{C\text{-product}}{CH_3.CO.OCH_3}$$

$$\xrightarrow{\text{(Cl}_3\text{C)}_2\text{CO}} \underset{e\text{-product}}{CH_3Cl} \tag{40}$$

The radical-pair precursor is a singlet; $\Delta g = g_A - g_B$, and owing to the presence of oxygen[3] in radical B, $g_B > g_A$ and Δg is negative; a_i is known to be negative in the methyl radical. Thus for methyl acetate, a c-product,

$$\Gamma_{NE} = - + - - = - \equiv E$$

For methyl chloride, an e-product,

$$\Gamma_{NE} = - - - - = + \equiv A^4$$

(2) A singlet pair of t-butyl radicals produced by peroxide decomposition disproportionate to yield isobutane and isobutene [equation (41)]. Both products show E/A multiplet effects.

$$\overline{2(CH_3)_3C\text{·}}^{\text{S}} \rightarrow HC(CH_3)_3 + H_2C = C(CH_3)_2 \tag{41}$$

[3] For qualitative purposes it is useful to remember that, owing to spin–orbit coupling, the radical in the radical pair containing a heavy atom (i.e. atom with higher mass number than carbon) will have the larger g-value. This effect has been thoroughly documented in CIDNP studies of [1]H (Closs and Trifunac, 1970a, b) and [19]F (Bethell et al., 1972b).

[4] In using Kaptein's rules in later sections, the sequence of signs will be that corresponding to the parameters as given in equations (39a) and (39b).

For the t-butyl radical a_H is positive[5] and necessarily $\Delta g = 0$. Since in isobutane J_{ij} is positive and the methine proton came originally from the other radical (i.e. σ_{ij} is negative),

$$\Gamma_{ME} = - + + + + - = + \equiv E/A$$

For isobutene, J_{ij} is negative and the splitting is due to protons on the same fragment; therefore

$$\Gamma_{ME} = - + + + - + = + \equiv E/A$$

It can be seen how useful the Kaptein rules are for analysing spectra quickly. However, there are several warnings to be given about their use: (i) a small multiplet effect superimposed on a large net effect can invalidate the rules; (ii) point (5) above must be considered in analysing the CIDNP spectra of magnetically equivalent nuclei; (iii) the rules assume T_0–S mixing, which is usually the case if the experiment is carried out in the probe of a high resolution n.m.r. spectrometer (see, however, Section IV).

Some of the uncertainties and ambiguities of qualitative rules such as Kaptein's and Müller's (1972) can be cleared up by computer simulation of polarized spectra. Alternatively, such quantitative approaches can be used to deduce a- and Δg-values.

IV. CIDNP AT LOW MAGNETIC FIELDS

Only a few workers have studied the dependence of CIDNP spectra on the magnetic field (Lehnig and Fischer, 1969, 1971a; Ward et al., 1969c; Garst et al., 1970; Garst and Cox, 1970; Slonim et al., 1970; Rykov et al., 1969b; Kaptein, 1971b, Kaptein and den Hollander, 1972). The basic problem is that at low magnetic fields all three components of the T state mix with the S state and as mentioned earlier $T_{\pm 1}$–S mixing involves nuclear spin flipping. Thus e- and c-products have the same sign of polarization. While this reduces the amount of derivable chemical information, there are redeeming qualities to low-field studies in that one can study the behaviour and effect of the sign of the exchange integral J_{ee} or ΔE_{ST}.

In low magnetic fields (less than 100 G) the effect of Δg is small and therefore transition probabilities between the α and β nuclear spin

[5] The hyperfine coupling constant $a_{\alpha-H}$ is negative and $a_{\beta-H}$ is positive in almost all aliphatic organic radicals.

states are approximately equal. It is therefore not necessary to consider polarization from T_0–S mixing.

For $T_{\pm 1}$–S mixing in a system with only one proton, $S(\alpha) \to T_{+1}(\beta)$ and $S(\beta) \to T_{-1}(\alpha)$ are the only allowed transitions. The energy separation (and hence the transition probability) between these states depends critically upon the values of J_{ee}, a_H, and the magnetic field (H_r) in which the reaction is conducted. Qualitative conclusions concerning the effects of $T_{\pm 1}$–S mixing on the observed polarization, applicable to a single group of nuclei, can be summarized as follows:

(1) S-precursors give opposite polarization to T and F precursors;
(2) c- and e-products show the same polarization;
(3) for an S precursor, if $J_{ee} < 0$, negative a_H gives polarization A; positive a_H may give either A or E polarization depending on $|J_{ee}|$ and H_r;
(4) if $J_{ee} > 0$ (S state lower in energy than the T state) then conclusion (3) is reversed; positive a_H gives E, negative a_H gives polarization which is a function of $|J_{ee}|$ and H_r.

As will be shown in Section V, $T_{\pm 1}$–S mixing has so far been identified in reactions involving diradicals and in certain reaction of organometallic compounds.

The zero field case (actually in earth's magnetic field of about 0·5 G) yields slightly different results (Kaptein, 1971b; Kaptein and den Hollander, 1972, see also, Charlton and Bargon, 1971; Ward et al., 1969c). No polarization can arise for a single nucleus since there is no preferred direction of quantization. If nuclear–nuclear spin coupling is operative, populations of nuclear Zeeman levels are unequal at zero-field. The effect in the CIDNP spectrum is that one transition is missing in each multiplet, due to the fact that certain transitions are between equally populated states. Kaptein has formulated a rule for the zero-field case.

If the product contains two coupled groups of equivalent nuclei with spin $\frac{1}{2}$, n_i and n_j, then the normal n.m.r. spectrum would contain $n_i + 1$ and $n_j + 1$ lines for nuclei i and j, respectively.

The "$n - 1$ multiplet rule" states that nuclei i and j will exhibit only n_i and n_j lines. The two missing lines are the high field line of the low field multiplet and the low field line of the high field multiplet. In addition the intensity of emission and absorption over both multiplets balances as would be expected. For example, a high field E/A multiplet would at zero-field exhibit E and A "$n - 1$ multiplets" for nuclei i and j respectively.

V. Applications of CIDNP

A. Scope and Limitations

1. Factors

For the purpose of discussing the applications of CIDNP, the foregoing account of the origin of the phenomenon may be summarized as follows. The observation of CIDNP in a chemical reaction requires

(i) the formation of intermediate radical pairs with lifetimes ($>$ ca. 10^{-9} sec) sufficiently long for mixing of singlet and triplet states (usually T_0–S mixing) to occur;

(ii) magnetic interaction between the odd electrons and atomic nuclei within the components of the radical pair so that nuclear spin selection occurs in the singlet–triplet mixing process;

(iii) the formation of chemically different products from singlet and triplet radical pairs;

(iv) appropriate experimental arrangements (magnetic field, reaction temperature, etc.) so that the phenomenon can be conveniently observed.

If these requirements are fulfilled, then the pattern of polarized signals observed in the n.m.r. spectrum of the particular reaction product will depend on the following additional factors:

(v) the multiplicity of the radical pair formed initially (usually taken to be the same as its precursor);

(vi) the mode of formation of the product from the radical pair, whether by combination of the component radicals ("cage product") or by reaction of one of them with another, non-radical species ("escape product");

(vii) the relative magnitude of the g-values of the components of the radical pair;

(viii) the signs and magnitudes of the hyperfine splitting constants for the nuclei observed;

(ix) the signs of nuclear–nuclear coupling constants;

(x) the magnetic field strength, which determines the relative importance of the Δg-effect (net effect) and hyperfine interactions (multiplet effect) in the T_0–S mixing process and may in some cases determine the relative importance of $T_{\pm 1}$–S and T_0–S mixing;

(xi) the values of the relaxation times of nuclear spin states in the reaction product under the reaction conditions.

Studies of CIDNP are thus potentially capable of providing information on any one of these factors provided that due caution is exercised. For an unequivocal conclusion to be reached concerning a particular factor, knowledge of the other determining factors is required, but often this need only be of a qualitative or semi-quantitative nature. Such applications of CIDNP are reviewed in general terms in the present section, and the literature up to the end of 1972 is surveyed, the material being arranged according to the type of reactant.

2. *Experimental methods*

One of the most attractive features of CIDNP studies is the ease with which experiments can be carried out. The basic requirement is only a high-resolution n.m.r. spectrometer.

For thermal reactions a variable temperature probe is necessary since optimum polarized spectra are usually obtained in reactions having a half-life for radical formation in the range 1–5 minutes. Reactant concentrations are usually in the range normally used in n.m.r. spectroscopy, although the enhancement of intensity in the polarized spectrum means that CIDNP can be detected at much lower concentrations. Accumulation of spectra from rapid repetitive scans can sometimes be valuable in detecting weak signals.

For photochemical studies, some workers recommend modification of the probe so that light can be passed through the (quartz) sample tube in a transverse direction (e.g. Blank *et al.*, 1971; Fischer, personal communication). Others have described methods of introducing light vertically downwards using a system of lenses and a mirror in conjunction with a light guide, for which a length of quartz rod passing from the mouth of the tube to below the surface of the sample serves quite well (Tomkiewicz and Klein, 1972; but see also, Maruyama *et al.*, 1971a). Powerful light sources (~500 W) are necessary to ensure a detectable signal. Great care is necessary to ensure that the photochemical reaction takes place in the sample region sensed by the n.m.r. probe.

3. *Deductions*

(a) *The observation of polarization.* Clearly, if polarized n.m.r. signals are observed in the products of a chemical reaction during and shortly after the reaction period, requirements (i)–(iv) of Section V,A,1 have been fulfilled. Thus, by reversing the logic, CIDNP can be used for the detection of stepwise radical mechanisms in chemical transformations and as such it is a very sensitive technique. There is at present no simple theoretical way by which the spectral enhancement may be accurately predicted for a particular reaction. Empirically, enhancement factors in the range 10^2–10^3 are typical, but both higher and lower values are possible.

In most cases CIDNP is substantially more sensitive in detecting products produced by way of radical-pair intermediates than e.s.r. spectroscopy is in detecting free radicals. No hard and fast general rules are possible, however.

This same sensitivity can, however, be misleading. Even minor radical routes to a particular reaction product could give rise to intense polarized n.m.r. signals which could obscure the normal monotonic increase of the signal due to product formed by a non-radical route. This problem can be overcome in some cases by estimation of the spectral enhancement factor. Again, it is not possible to justify a firm, threshold value, but as a useful "rule of thumb" when enhancements fall below about 100 then the possibility of an important alternative non-radical route to the same product should be carefully investigated.

Besides providing evidence of radical intermediates, CIDNP can also provide useful evidence concerning the formation of minor products since these may appear very much more prominently in the polarized spectrum during reaction than in the depolarized spectrum when reaction is complete. Metastable products may also be detected by the observation of a polarized spectrum during reaction which differs from the final depolarized spectrum (see for example Bethell *et al.*, 1972c). Further evidence on this point can sometimes be obtained by studying the kinetics of the appearance and decay of the polarized signals under the reaction conditions and comparing the results with those obtained by conventional determination of relaxation times. Such experiments should also assist in differentiating true polarized signals from signals due to transient *unpolarized* intermediates.

Failure to observe polarization in a particular reaction is significant only to the extent that any negative evidence is significant. If other evidence points to a radical pathway for the reaction, it may well be worth checking that the nuclear relaxation times for nuclei in the product are not unexpectedly short and also that polarization is not observable in a different spectral region from that expected for the final product owing to the formation of a metastable intermediate.

(b) *The pattern of polarization.* The pattern of polarization in a spectrum displaying CIDNP provides information concerning factors (v)–(xi) of Section V,A,1. With the exception of the magnetic field-dependence and the effects of relaxation times, these factors are conveniently embodied in a qualitative way in Kaptein's rules (Section III).

In some instances, however, where all the parameters have been known if only qualitatively, the pattern of polarization has turned out to be different from that predicted theoretically. An explanation has therefore been advanced in terms of interconversion of radical pairs. This can be explained using Scheme 1. Suppose a singlet precursor species gives

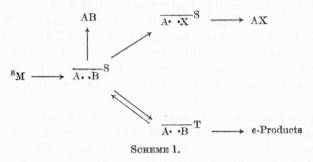

SCHEME 1.

rise to a radical pair $\overline{A \cdot \cdot B}^S$ and this, by T_0–S mixing, leads to polarization in the c-product AB and the e-products. The patterns of polarization in the products could then be deduced from the parameters appropriate to $\overline{A \cdot \cdot B}^S$. If on the other hand a transformation of $\overline{A \cdot \cdot B}^S$ into a second radical pair $\overline{A \cdot \cdot X}^S$ occurred and the formation and subsequent reaction of this second pair were rapid compared with the time required for it to undergo T_0–S mixing, then the product AX would have its nuclei polarized in a manner determined by the parameters appropriate to $\overline{A \cdot \cdot B}^S$. Such transformations may take place, for example, by loss of carbon dioxide from an acyloxy radical $R.CO.O\cdot$ or by a radical transfer reaction with a solvent molecule (the so-called "cage wall effect" or radical-pair substitution). Inevitably, the term "memory effect" has been coined to describe the carrying over of polarization from one radical pair to another.

The effect has been most commonly encountered in the decomposition of symmetrical diacyl peroxides where it is easily recognized since the symmetrical radical dimer, for which Δg must be zero, is formed and shows *net* polarization. Clearly, studies of such systems are capable of providing valuable information on the dynamics of radicals and radical pairs in solution, the polarization process providing a time base for events (see Section V,B).

The predictions of simple rules such as Kaptein's and Müller's can be distorted by relaxation effects. These are particularly noticeable in photochemical experiments. In the pre-steady state (e.g., immediately after irradiation has begun), when build-up of the polarized signals is occurring, relaxation effects in the final product are relatively unimportant and observed spectra accord with the simple theory. Conversely, immediately after the irradiation is stopped, relaxation effects are paramount; changes in relative line intensities and even inversion of the phase of the polarized signal can then be observed (Müller and Closs, 1972). Methods for computer simulation of such relaxation effects have been developed (Rykov et al., 1970b; Müller and Closs, 1972).

The simple rules refer to nuclear spin selection during mixing of T_0 and S states of the radical pair. If mixing of $T_{\pm 1}$ and S states occurs instead, entirely different rules apply as discussed in Section IV (Kaptein, 1971b, 1972a; Kaptein and den Hollander, 1972; Garst et al., 1971; J. I. Morris et al., 1972).

Other possibilities (e.g., Overhauser effects) exist for the complication of the patterns of polarization predicted by the simple theory. Interesting transient splittings have been observed in polarized ^{19}F-spectra (Bethell et al., 1972a, b). Nevertheless, straightforward application of the simple rules seems to yield reliable conclusions in most cases. Clearly, however, it is unwise to rely on CIDNP results alone in studies of organic reaction mechanisms; other information is invariably necessary to ensure that the correct interpretation is chosen from among the several possibilities which CIDNP may suggest.

B. *Reactions of Diacyl Peroxides and Related Compounds*

1. *Introduction*

Diacyl peroxides undergo thermal and photochemical decomposition to give radical intermediates (for a recent review, see Hiatt, 1971). Mechanistically the reactions are well understood as a result of the many investigations of products and kinetics of thermal decomposition (reviewed by DeTar, 1967; Cubbon, 1970). Not surprisingly, therefore, one of the earliest reports of CIDNP concerned the thermal decomposition of benzoyl peroxide (Bargon et al., 1967; Bargon and Fischer, 1967) and peroxide decompositions have been used more widely than any other class of reaction in testing theories of the phenomenon.

Current views on the main features of the mechanisms of decomposition of benzoyl peroxide are summarized in Scheme 2.

The kinetic form of the decomposition in various solvents indicates competing unimolecular homolysis of the peroxide link (a) and radical induced decomposition (b). Other diacyl peroxides behave similarly, except that, in the case of acetyl peroxide, induced decomposition is much less important. More highly branched aliphatic or α-phenyl-substituted diacyl peroxides decompose more readily, partly because induced decomposition is more important again and partly because of the occurrence of decomposition involving cleavage of more than one bond (for a mechanistic discussion of these cases, see Walling et al., 1970).

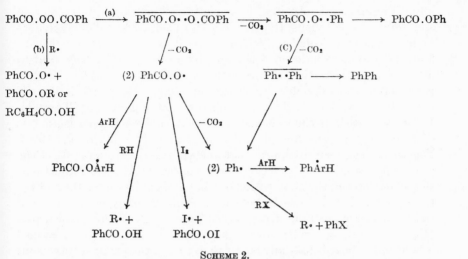

SCHEME 2.

For benzoyl and acetyl peroxides, loss of carbon dioxide occurs in a stepwise process. Estimates of the rate constants for step c in Scheme 1 are 7×10^3 sec^{-1} (benzene, 60°). The corresponding process for acetyl peroxide has $k = 2 \times 10^9$ sec^{-1} (n-hexane, 60°), so that the lifetime of radical pairs containing acetoxy radicals is comparable to the time necessary for nuclear polarization to take place (Kaptein, 1971b; Kaptein and den Hollander, 1972; Kaptein et al., 1972). Propionoxy radicals are claimed to decarboxylate 15–20 times faster than acetoxy radicals (Dombchik, 1969).

CIDNP studies of the decomposition have centred mainly on thermal decompositions; photochemical decomposition has generally been less intensively investigated. While most reports of polarization refer to ^{1}H n.m.r. spectra, a number of papers have described polarization of other nuclei, ^{2}H (Kaptein, 1971b; Kaptein et al., 1972), ^{13}C (Lippmaa et al., 1970a, b, 1971; Kaptein, 1971b; Kaptein et al., 1972; Kessenikh et al., 1971), and ^{19}F (Kobrina et al., 1972) contained in the peroxide reactant. Additionally, polarization of ^{31}P has been reported in the products of decomposition of benzoyl peroxide in phosphorus-containing solvents (Levin et al., 1970).

2. Decomposition of diacyl peroxides in non-aromatic solvents

(a) *Patterns of polarization.* Thermal decomposition of benzoyl peroxide in a variety of organic solvents leads to the production of a strong emission signal in the ^{1}H n.m.r. spectrum attributable to benzene. The signal is strongest in cyclohexanone and cyclohexane; a lower-intensity emission is observed in solvents such as carbon tetrachloride and tetrachloroethylene (Rykov and Buchachenko, 1969). Solvents such as isopropyl alcohol and ethylene glycol monoethyl ether which are known to promote induced decomposition of benzoyl peroxide lead to little or no polarization (Fahrenholtz and Trozzolo, 1971). No ^{1}H-polarization is observed when perdeuteriobenzoyl peroxide is decomposed in a suitable ^{1}H-donating solvent (Fahrenholtz and Trozzolo, 1972), and with perfluorobenzoyl peroxide only ^{19}F polarization is detected in the pentafluorobenzene produced (Kobrina et al., 1972). Together, these results confirm that nuclear spin selection occurs at the radical-pair stage of the reaction. To be consistent with the radical-pair theory of CIDNP, the observation of a net emission requires that, of the three radical pairs shown in Scheme 2, benzene must be formed in a radical transfer reaction of the singlet benzoyloxy-phenyl radical pair ($\Gamma_{NE} = - - - + = - \equiv E$), since the g-value of the benzoyloxy radical is believed to be substantially larger than that of the phenyl component. Consistently, irradiation of benzoyl peroxide in cyclohexane containing a triplet sensitizer, for example acetophenone-d_5, produces a benzene signal showing enhanced absorption (Fahrenholtz and Trozzolo, 1971). Similar results are observed with other sensitizers, provided that their triplet energy is greater than about 56 kcal mol^{-1} (cf. Walling and Gibian, 1965) and that they are not capable of singlet energy transfer to the peroxide. Thus benzil ($E_T = 54$ kcal mol^{-1}) fails to sensitize decomposition of the peroxide which occurs instead by direct photolysis giving an emission signal from benzene, and triphenylene-d_{12} ($E_T = 67$ kcal mol^{-1}; fluorescence quantum yield 0·11, indicating relatively inefficient intersystem crossing to the excited triplet state compared with acetophenone) also leads to the observation of emission. Analogous changes in the phase of polarization brought about by sensitized photolysis of benzoyl peroxide in carbon tetrachloride have also been reported (Kaptein et al., 1970; Fahrenholtz and Trozzolo, 1971; see also Lehnig and Fischer, 1969).

Since the benzene emission in the thermal decomposition of benzoyl peroxide results from radical transfer by the phenyl component of a benzoyloxy-phenyl radical pair, phenyl benzoate produced by radical combination within the same pair should appear in absorption. A weak transient absorption has been tentatively ascribed to the ester (Lehnig and Fischer, 1970) but the complexity of the spectrum and short relaxation time (Fischer, personal communication) makes unambiguous assignment difficult. Using 4-chlorobenzoyl peroxide in hexachloroacetone as solvent, however, the simpler spectrum of 4-chlorophenyl 4-chlorobenzoate is clearly seen as enhanced absorption, together with the strong emission of p-dichlorobenzene (Blank and Fischer, 1971a, b). Analogous results in the thermal decomposition of acetylbenzoyl peroxide have been reported (Buchachenko et al., 1970a).

Polarization in the ^{13}C n.m.r. spectrum (15·1 MHz) of decomposing solutions of non-enriched benzoyl peroxide in cyclohexanone (110°C) are more revealing (Lippmaa et al., 1970a, b; see also Schulman et al., 1972). Assignment of peaks is simplified by comparing spectra obtained with and without decoupling of protons, since this permits resonances due to carbon atoms that bear hydrogen to be distinguished from those of carbon atoms that bear hydrogen to be distinguished from those of carbon atoms that do not. The strongest of the polarized transitions in the ^{13}C-spectrum are attributable to phenyl benzoate (C-1 of the two phenyl groups and the carbonyl carbon atom: all A). Benzene appears in emission as a doublet, collapsing to a singlet in the double resonance experiment. In addition, carbon dioxide appears in emission, and carbon atoms 1 and 1′ of biphenyl (a very minor product arising from a secondary radical pair) as a weakly enhanced absorption. ^{13}C-Polarization is apparently larger than ^1H-polarization and occurs only for carbon atoms close to the radical centre. With reasonable assumptions about the hyperfine ^{13}C splitting constants in phenyl and benzoyloxy radicals, the pattern of polarizations is readily interpretable in terms of the radical pair theory (Fischer, personal communication).

The polarization of biphenyl, deserves special comment. If, as indicated in Scheme 2, its immediate precursor is a radical pair consisting of two phenyl radicals, then it should be formed without detectable net polarization since if $\Delta g = 0$. Analogous results have been reported in the decomposition of other peroxides; for example, ethane formed from acetyl peroxide shows net emission. To account for this, it has been suggested (Kaptein, 1971b, 1972b; Kaptein et al., 1972) that nuclear spin selection which occurs in the primary radical pair—in the present instance, $\overline{PhCO.O \cdot \cdot Ph}^S$—is carried over into the derived radical pair $\overline{Ph \cdot \cdot Ph}^S$ and thence to the recombination product. The biphenyl is to be regarded therefore as if it were a recombination product of the initial radical pair. The magnitude of this "memory effect" can be related to the velocity constant for the transformation of the unsymmetrical radical pair into the symmetrical pair. It is consistent with this picture that, in aliphatic diacyl peroxides, such as dipropionyl peroxide, only multiplet polarization is detected, indicating nuclear spin selection in ethyl radical pairs with $\Delta g = 0$ (E/A in butane; A/E in ethyl iodide formed in the presence of iodine or alkyl iodide). The lifetime of acyloxy-alkyl radical pairs is too short for spin selection at this stage so that the combination product ethyl propionate shows no polarization (Cooper *et al.*, 1972a).

In the decomposition of benzoyl peroxide, the fate of benzoyloxy radicals escaping from polarizing primary pairs remains something of a mystery. Benzoic acid is formed but shows no polarization in [1]H- and ^{13}C-spectra, and the carboxylic acid produced in other peroxide decompositions behaves similarly (Kaptein, 1971b; Kaptein *et al.*, 1972). Some light is shed on the problem by studies of the thermal decomposition of 4-chlorobenzoyl peroxide in hexachloroacetone containing iodine as

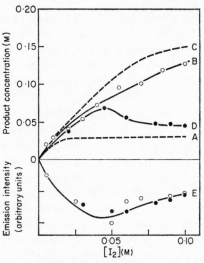

FIG. 8. Product yields and polarization (100 MHz) in the thermal decomposition (125°C) of 4-chlorobenzoyl peroxide (0·1 M) in hexachloroacetone containing iodine and water. Curve A, 4-chlorobenzoic acid ([H_2O], *ca.* 0·03 M); Curve B, *p*-chloroiodobenzene ([H_2O], *ca.* 0·03 M); Curve C; 4-chlorobenzoic acid ([H_2O], *ca.* 1·6 M); Curve D, *p*-chloroiodobenzene ([H_2O], *ca.* 1·6 M); Curve E, maximum intensity of emission from *p*-chloroiodobenzene (○, *ca.* 0·03 M H_2O; ●, *ca.* 1·6 M H_2O). Data of Blank and Fischer, 1971b.

scavenger and a little water (Blank and Fischer, 1971a, b). Iodine can trap aroyloxy radicals as the hypoiodite, $ArCO.OI$. In anhydrous conditions this decarboxylates giving ArI, but it is readily hydrolysed by water. The effect of iodine and water on the products of the peroxide decomposition is shown in Fig. 8. The amount of 4-chlorobenzoic acid increases monotonically, while the *p*-chloroiodobenzene yield rises to a maximum and then falls; with low water concentrations the latter product shows a monotonic increase. However, only *p*-chloroiodobenzene appears polarized (E) in the [1]H n.m.r. spectrum and, irrespective of the water concentration, polarization increases to a maximum and then falls away as the iodine concentration increases. It thus appears that polarization is not observed in *p*-chloroiodobenzene produced by way of the hypoiodite, suggesting that nuclear spin relaxation in this intermediate may be very rapid. Spin-lattice relaxation in the free acid is also probably very rapid. Indeed, the relaxation time for 4-chlorophenyl 4-chlorobenzoate is only about one-eighth of that of *p*-chloroiodobenzene (Blank and Fischer, 1971b). However, both relaxation times are approximately halved in the presence of 0·1 M iodine so that several alternative interpretations of the results in Fig. 8 become possible.

(b) *Scavenging experiments.* Use of scavengers other than iodine in conjunction with CIDNP studies has provided valuable evidence concerning the dynamics of radical pairs in solution. Figure 9 shows the

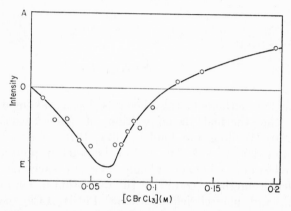

Fig. 9. Intensity of maximum polarization (100 MHz) of chloroform produced in the thermolysis (80°C) of isobutyryl peroxide (0·2 M) in hexachlorobutadiene containing bromotrichloromethane. Data of Kaptein *et al.*, 1971.

effect of the concentration of added bromotrichloromethane on the polarization of chloroform produced in the thermal decomposition of isobutyryl peroxide in hexachlorobutadiene (Kaptein *et al.*, 1971; Kaptein, 1971b, 1972b). The emission observed at low concentrations of CCl_3Br must imply hydrogen abstraction by trichloromethyl radicals ($g = 2·0091$) from isopropyl radicals ($g = 2·0026$) (Scheme 3a) and, since $a_{\beta-H}$ is positive, this indicates that the polarization must occur in a radical pair formed by diffusive encounter of free radicals ($\Gamma_{NE} = + + - + = - \equiv E$). The change to enhanced absorption at higher concentrations of CCl_3Br therefore shows that the precursor of chloroform must change to one having singlet character. This can be

4

understood if the symmetrical isopropyl-isopropyl radical pair produced in the primary, multi-bond homolysis reacts with the scavenger to produce a new isopropyl-trichloromethyl radical pair by radical-pair substitution as shown in Scheme 3,b (Cooper *et al.*, 1972a, b).

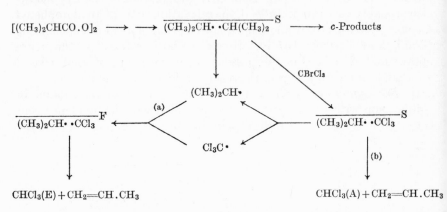

SCHEME 3.

Bromotrichloromethane has also been used as a scavenger in CIDNP studies on the thermal decomposition of phenylacetyl peroxide $(PhCH_2CO.O)_2$ (Walling and Lepley, 1971, 1972). In the absence of scavenger, no polarization is detected in the reaction products, namely benzyl phenylacetate, bibenzyl and benzyl benzoyl carbonate. Addition of bromotrichloromethane results in the formation of benzyl bromide, polarized E, and phenyltrichloroethane, $PhCH_2.CCl_3$, polarized A. Likewise, benzyl chloride and phenyltrichloroethane are formed with the scavenger Cl_3CSO_2Cl. The phases of the polarized signals are consistent with phenyltrichloroethane being the result of collapse of a benzyl-trichloromethyl radical pair formed by diffusive encounter $(\Gamma_{NE} = + + - - = + \equiv A)$, and the benzyl halides would arise by reaction of the benzyl component with a further molecule of scavenger $(\Gamma_{NE} = + - - - = - \equiv E)$. The polarization of both products, however, shows pronounced changes with scavenger concentration, the intensity passing through a maximum, although there is no change in phase as in the case of isobutyryl peroxide. The changes for the benzyl halides are quantitatively interpretable in terms of spin selection in the radical pairs, the polarization increasing with the number of diffusive encounters of the free radicals and decreasing as a result of the usual relaxation processes in the free radical. Derived values for velocity constants for these rate processes are reasonable. However, this simple scheme plainly will not explain the observations for phenyltrichloroethane since it cannot build up polarization by successive polarizing encounters. An alternative interpretation of the variation in polarization in both products could be the incursion of radical-pair substitution at high scavenger concentrations.

Further light on the process of radical-pair substitution comes from detailed studies of the scavenging action of alkyl iodides on the thermal decomposition of benzoyl peroxide (Ward *et al.*, 1969a; Rykov *et al.*, 1970a; Ignatenko *et al.*, 1971; Cooper *et al.*, 1972a, b). Using ethyl iodide, for example, the principal polarizations are of iodobenzene (E), ethyl benzoate (CH$_2$, A; CH$_3$, E); ethylbenzene (CH$_2$, E + E/A) and ethyl iodide[6] (CH$_2$, A + E/A; CH$_3$, E + E/A). The ethyl benzoate net polarization requires the precursor to be a geminate benzoyloxy-ethyl radical pair,[7] again the result of radical-pair substitution in a benzoyloxy-phenyl pair with expulsion of polarized iodobenzene. The E + E/A pattern of polarization of ethylbenzene indicates contributions from two precursors of opposite Δg; the benzoyloxy-ethyl pair leads to E and the E/A component arises from the derived phenyl-ethyl pair. Significantly, the pattern of polarization of ethyl benzoate and ethylbenzene produced in the thermolysis of benzoyl propionyl peroxide (without added scavenger) is exactly the same, thus supporting this interpretation. The dependence of iodobenzene and ethyl benzoate polarization on the scavenger concentration is similar to that of benzyl halide in the experiments with phenyl alkyl peroxide. At low ethyl iodide concentrations, iodobenzene shows strong emission, increasing rapidly, as the availability of the scavenger permits trapping of phenyl radicals before nuclear spin relaxation can occur ($< 10^{-4}$ sec). At 0·4 M ethyl iodide, however, trapping becomes so efficient that there is insufficient time for the primary benzoyloxy-phenyl pair to undergo increased nuclear spin selection. Thereafter, the iodobenzene polarization falls. Concomitantly, ethyl benzoate polarization becomes detectable and reaches a maximum when the ethyl iodide concentration is so high that essentially all the benzoyloxy-phenyl pairs are trapped. The qualitative conclusions are embodied in Scheme 4.

Clearly, further investigation is desirable before these observed variations in polarization with scavenger concentration can be unequivocally interpreted. However, these studies hold out high promise not only of

---

[6] Polarized signals due to the scavenger when allyl iodide is used are broadened, probably as a result of the exchange process CH$_2$=CH.CH$_2$I $\rightarrow$ [CH$_2$⋯CH⋯CH$_2$]$^{\cdot}$ $\rightarrow$ ICH$_2$.CH=CH$_2$ (Lawler *et al.*, 1971).

[7] The same radical pair gives rise to ethylene polarized A as a result of the combination of α-protons from the ethyl radical (polarized E) and a contribution from the β-protons (A) which is dominant on account of their larger hyperfine coupling constant.

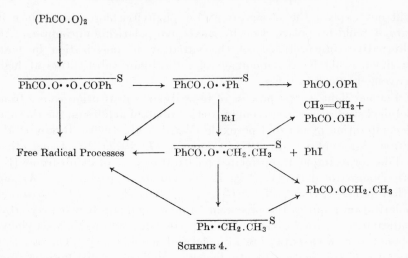

SCHEME 4.

yielding qualitative pictures of the processes involved, but also of providing quantitative information on the dynamics of radical-pair and free radical reactions.

(c) *Miscellaneous peroxide decompositions.* In addition to thermolysis and photolysis of diaroyl, dialkanoyl, and mixed aroyl alkanoyl peroxides, CIDNP has also been observed in thermolyses of related peroxy-compounds. Thus simple dialkyl peroxides have been used (Kaptein, 1968, 1971b; Levin *et al.*, 1970). Peresters such as peracetyl isopropyl carbonate (Rykov *et al.*, 1969a, 1970a; Buchachenko *et al.*, 1970c) and di-t-butyl peroxalate (Ignatenko *et al.*, 1971) have been widely investigated, especially by Russian groups. Polarization of the products of scavenging experiments using alkyl iodides can be interpreted in terms of precursor radical pairs of low Δg (alkyl-alkyl) and high Δg (alkyl-alkoxy) (see, for example, Ignatenko *et al.*, 1971). In the thermolysis of t-butyl perpivalate, $(CH_3)_3C.CO.OOC(CH_3)_3$, di-t-butyl ether appears in emission (Kaptein, 1971b); it has been claimed that if this product arises directly from a t-butyl–t-butoxy radical pair derived very rapidly from the primary pair from the homolysis, then Δg would seem to be positive, i.e. $g_{t-BuO}. < g_{t-Bu}..$ This unexpected conclusion, while it can be rationalized, depends on the assumption that the hyperfine splitting constant for the protons in the t-butoxy radical is of the same sign as in the t-butyl radical; since there are difficulties in observing the t-butoxy radical in solution, independent evidence on this point is not available.

Kaptein (1971b; Kaptein *et al.*, 1972) has investigated a wide range of dialkanoyl peroxides by CIDNP and has concluded that the rate of decarboxylation of radicals $RCO.O\cdot$ decreases in the order $R = alkyl$ or cycloalkyl $> CH_3 \longrightarrow >$ isopropenyl $>$ cyclopropyl $> CH_2\!\!=\!\!CH\!\!-\!\!$ $> CH_3.CH\!\!=\!\!CH\!\!-\!\!$. For secondary or tertiary or resonance-stabilized alkyl radicals $R\cdot$, the acyloxy radical, if it is formed at all, probably does not survive outside the primary geminate pair. Particularly interesting results come from a study of cyclopropylacetyl peroxide $(c\text{-}C_3H_5.CH_2CO.O)_2$ (Kaptein, 1971b, 1972b). Thermolysis in hexachloroacetone gives, among other products, the rearranged transfer product 4-chloro-1-butene, $ClCH_2.CH_2.CH\!\!=\!\!CH_2$, in which the strongest polarization is observed for the vinylic CH_2. Since the corresponding a_H-value is very small in the butenyl radical, the polarization must arise from spin selection in the cyclopropylmethyl radical. It is therefore possible to deduce a lower limit for the velocity constant for the cyclopropylmethyl $\rightarrow$ butenyl rearrangement step of 3×10^7 sec^{-1}. Moreover, the CIDNP studies show that the vinylic proton on C-2 is correlated only with the vinylic protons on the adjacent carbon atom in the product. This excludes the rapid equilibration of the cyclopropylmethyl radical [equation (42)] as well as the "non-classical" bridged structure (1) since correlation would then be with all other protons. CIDNP studies thus promise to provide new insights into the subject of bridged radicals (see for example Hargis and Shevlin, 1973).

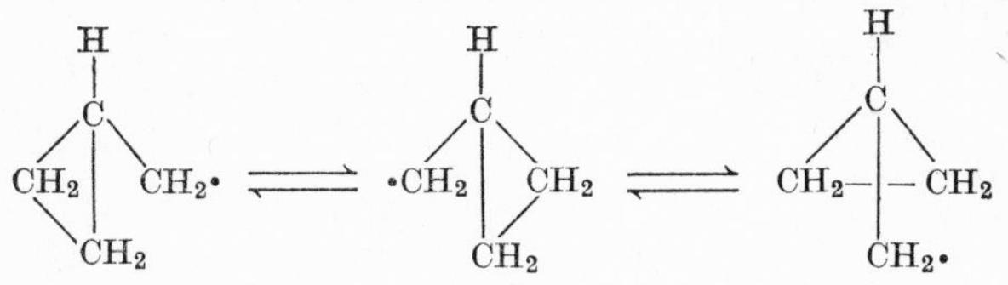

(42)

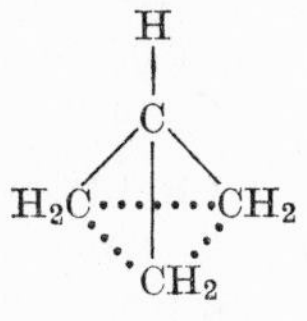

(1)

3. *Peroxide decomposition in aromatic and other unsaturated solvents: homolytic aromatic substitution and olefin polymerization*

Decomposition of peroxides in aromatic solvents leads to attack on the aromatic nucleus by radicals and hence to substitution products (for a recent summary, see Williams, 1970). In the substitution of benzene and related substrates by phenyl radicals, for example, cyclohexadienyl radicals (2) are thought to be intermediates since, under appropriate circumstances, these can be trapped by hydrogen atom donation or by dimerization. Recently, CIDNP evidence for their formation has been provided (Fahrenholtz and Trozzolo, 1972).

The n.m.r. spectra of typical reaction mixtures from homolytic aromatic substitution are very complex. To simplify the problem of spectral interpretation, perdeuteriobenzoyl peroxide has been used together with a symmetrically trisubstituted aromatic substrate, such as 1,3,5-trichlorobenzene. With these reactants in hexachloroacetone as solvent (127°), one observes single-line polarized signals which are due to 2,4,6-trichlorobiphenyl (A) and pentachloroacetone (E). Since radical-induced decomposition reactions, i.e. radical + molecule reactions, do not induce polarization, a radical-pair must be responsible and the sequence of processes shown in Scheme 5 satisfies the requirements.

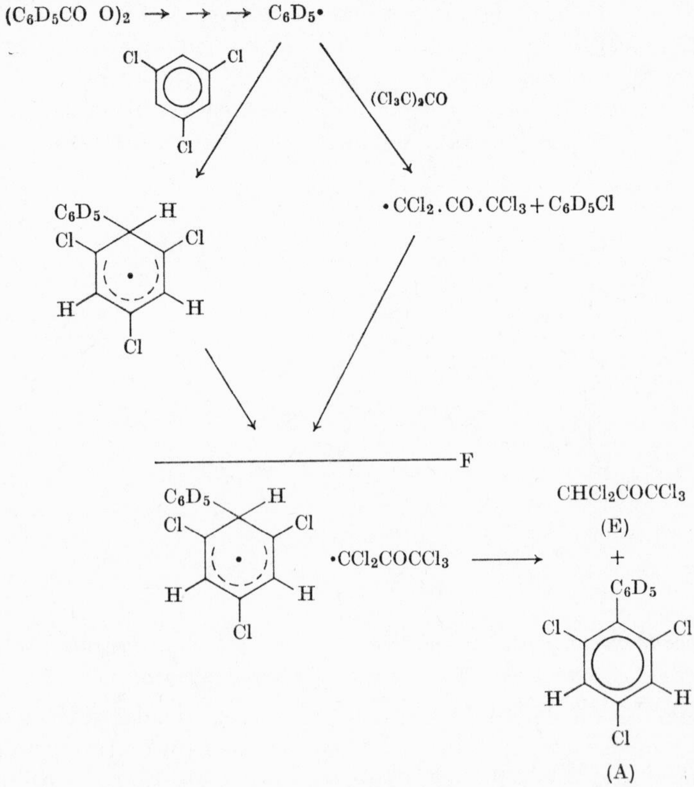

SCHEME 5.

The g-value of the pentachloroacetonyl radical is greater than that of the cyclohexadienyl radical and a_H is positive. Pentachloroacetone therefore appears in emission $(\Gamma_{NE} = + + - + = - \equiv E)$. In the absence of a_H-values for the vinylic protons in the cyclohexadienyl radical, no prediction can be made about the phase of the biphenyl polarization.

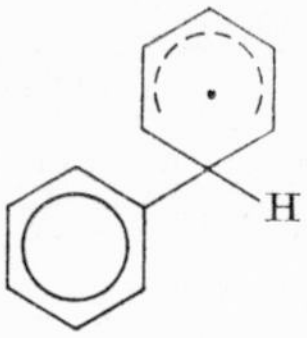

(2) 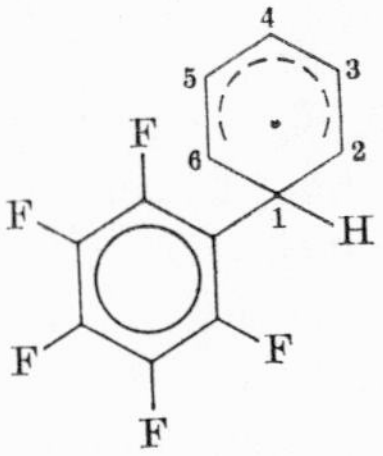   (3)

Using perfluorobenzoyl peroxide, radical attack on the aromatic ring occurs before decarboxylation of the perfluorobenzoyloxy radicals (Bargon, 1971b, c). It has been assumed that the attack occurs within the primary cage, that is to say, the reaction is a special case of radical pair substitution. The product of this step is a perfluorobenzoyloxy-perfluorophenylhexadienyl radical pair of which the former component is believed to have the larger g-value. The competing reactions of this pair are then reversion to reactants or disproportionation, giving perfluorobenzoic acid and substitution product. From e.s.r. studies on 3, the following a_H values have been assigned: $a_1 = +32\cdot4$ G; $a_2 = a_6 = -8\cdot8$ G; $a_3 = a_5 = +2\cdot75$ G and $a_4 = -12\cdot95$ G. Positive a-values predominate therefore, and in the case of benzene, formed by reversion of 3 reactants, emission is observed. However, in the product of substitution, the occurrence of absorption or emission depends on the sign of a, and this readily permits identification of the position of attack. It turns out that benzene containing an electron-releasing substituent (e.g., OCH_3) is attacked by the perfluorobenzoyloxy radical in the o- and p-positions while electron withdrawing substituents (e.g., NO_2) direct attack to the m-position. Clearly, perfluorobenzoyloxy radicals show the expected electrophilic character.

Few CIDNP studies on free radical reactions with olefins and related unsaturated molecules have been reported, and relatively little chemically useful information seems to have been derived, despite the potential relevance in polymerizing systems. Thus CIDNP has been reported in the decomposition of benzoyl peroxide in the presence of styrene and methyl methacrylate (Bargon and Fischer, 1968a, b) and in peroxide-initiated addition of tetrabromomethane to styrene (Fischer and Bargon, 1969; Bargon, 1971a). In the latter case, the ^{1}H n.m.r. spectrum shows perfect multiplet effects (E/A) for the aliphatic protons in the adduct PhCHBr.CH$_2$.CBr$_3$, and this has been interpreted as indicating diffusive encounter of two PhĊH.CH$_2$.CBr$_3$ radicals, formed by attack of tribromomethyl radicals on the double bond, which then abstract bromine from further molecules of CBr$_4$ or dimerize. However, polarization of the dimers has not been reported and further investigation is desirable. Polymerizing systems clearly deserve more detailed examination in the light of current theory. While the chain process involved in homopolymerization cannot alone give rise to polarization, CIDNP could well arise when certain chain termination processes occur, and this could provide important information.

It is convenient to refer here to the addition of free radicals, in this case generated from azo-bis-isobutyronitrile, to nitrones and nitroso-compounds. 1:2-Adducts are formed, as shown in equations (43) and

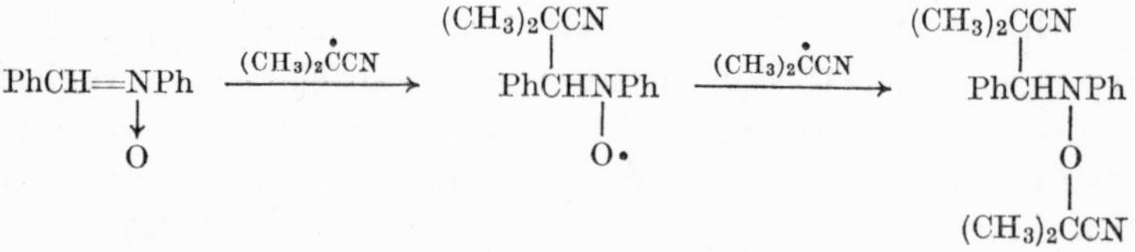

$$PhCH{=}NPh \xrightarrow{\text{(CH}_3)_2\dot{C}CN} \underset{\overset{|}{O\cdot}}{\overset{\text{(CH}_3)_2CCN}{\underset{|}{Ph\dot{C}HNPh}}} \xrightarrow{\text{(CH}_3)_2\dot{C}CN} \underset{\overset{|}{\overset{O}{\underset{|}{\text{(CH}_3)_2CCN}}}}{\overset{\text{(CH}_3)_2CCN}{\underset{|}{PhCHNPh}}} \quad (43)$$

$$PhN{=}O \xrightarrow{\text{(CH}_3)_2\dot{C}CN} \underset{\text{(CH}_3)_2CCN}{\overset{|}{PhN{-}O\cdot}} \xrightarrow{\text{(CH}_3)\dot{C}CN} \underset{\text{(CH}_3)_2CCN \quad CN}{\overset{|}{PhN{-}O{-}C(CH_3)_2}} \quad (44)$$

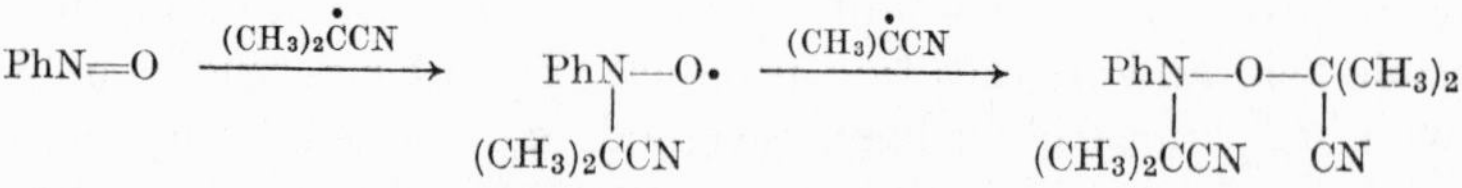

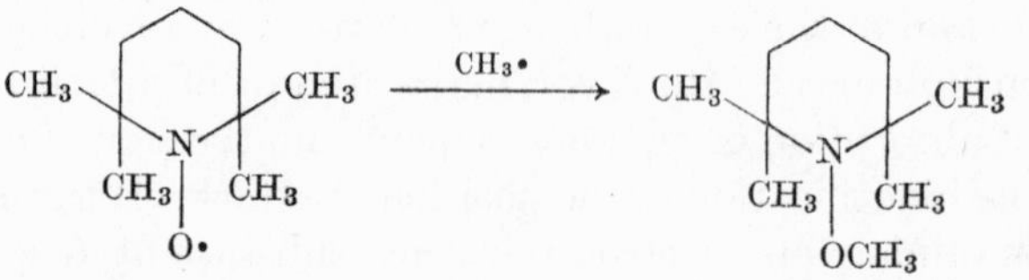

(44), but polarization (E) is observed only in the second step, namely the reaction of a second 2-cyano-2-propyl radical with the intermediate nitroxyl radical (Iwamura and Iwamura, 1970; Iwamura et al., 1970b; see also Iwamura et al., 1971b). This is consistent with the observation of polarization (A) of the O-methyl group in the adduct of methyl radicals (from acetyl benzoyl peroxide) with a stable nitroxyl radical [equation (45)] (Rykov and Sholle, 1971a, b).

C. *Reactions of Azo-, Diazo- and Related Compounds*

1. *Azo-compounds*

The thermolysis and photolysis of azo-compounds ($R.N:N.R$) are well documented sources of radicals (for a review, see Strausz *et al.*, 1972). As such they have been extensively used as radical sources in CIDNP studies (Fischer and Bargon, 1969; Closs and Trifunac, 1969, 1970b, c; Iwamura and Iwamura, 1970; Iwamura *et al.*, 1970b; Kasukhin *et al.*, 1970). The application of CIDNP to unravel some of the mechanistic complexities of the decomposition of azo-compounds has now begun.

A central problem in this area concerns the timing of the cleavage of the two carbon–nitrogen bonds. Synchronous and consecutive cleavage can be envisaged [equation (46)] and conventional studies of kinetics and

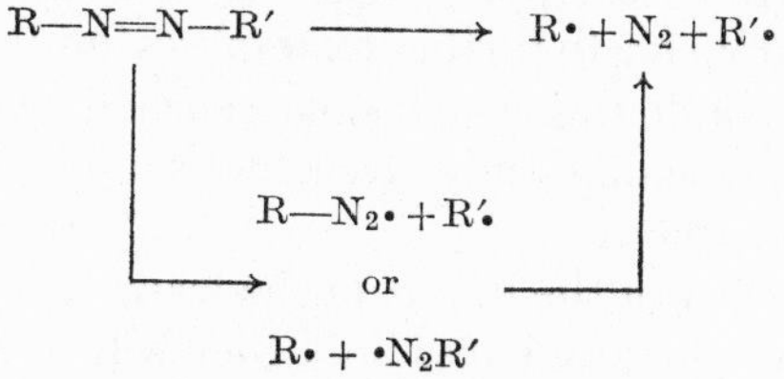

$$R\text{—}N\text{=}N\text{—}R' \longrightarrow R\cdot + N_2 + R'\cdot$$

$$R\text{—}N_2\cdot + R'\cdot \tag{46}$$

$$\text{or}$$

$$R\cdot + \cdot N_2R'$$

products suggest that under appropriate structural circumstances both routes are followed. Related to this problem, there is continuing interest in the dynamics of the radicals formed in the primary decomposition step, and its relation to spin correlation in radical pairs. CIDNP studies have illuminated both, as well as throwing new light on the mechanism of the photochemical decomposition of azo-compounds.

Photolysis at about 40°C of the *trans*-isomer of the unsymmetrical azo-compound 4 in benzene solution yields dicumyl and biphenyl as the major products, together with a little α-methylstyrene, cumene, and 2,2-diphenylpropane. All these products show polarization of their

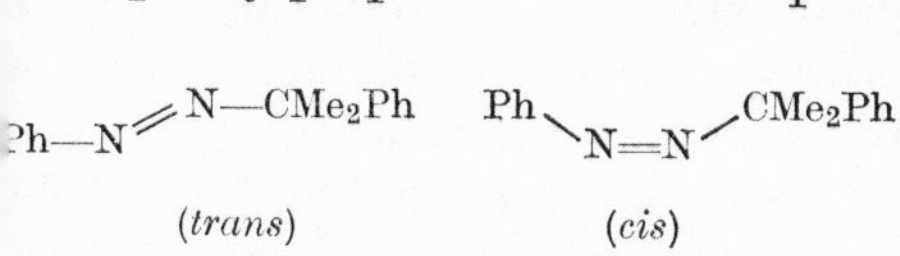

$$Ph\text{—}N{\overset{N\text{—}CMe_2Ph}{\diagdown}} \qquad \underset{N\text{=}N}{\overset{Ph}{\diagdown}}{\diagup}^{CMe_2Ph} \qquad Ph_2CH.N\text{=}N.CH_2Ph$$

(*trans*)   (*cis*)   (30)

(4)   (5)

H n.m.r. signals but, in addition, the starting material itself is polarized Porter *et al.*, 1972). Moreover, the appearance of polarization shows an induction period after irradiation is commenced and continues when the irradiation is interrupted for a time longer than the nuclear relaxation times of the products. The photochemical formation of a thermally labile intermediate, decomposition of which leads to radicals, is clearly indicated, and this intermediate has been identified as cis-4. This compound, on thermal decomposition at 25°C, yields a similar product mixture with the same polarization pattern including polarized trans-4. Thus, under these reaction conditions at least, the important photochemical reaction is merely a trans-cis isomerization of the azo-compound. This confirms reports on the low-temperature photolysis of simple trans-azoalkanes (e.g., Me_3C—N=N—CMe_3) which at $-50°$ yields only the cis-isomer (Mill and Stringham, 1969). Of necessity, photolysis of trans-4 leads eventually to radical pairs initially in the singlet state, a conclusion also reached for the azo-compound 5 on the basis of the pattern of polarization observed in the spectrum of 1,2,2-triphenyl-ethane, the correlated radical pair combination product (Closs and Trifunac, 1969, 1970b, c). Application of the CIDNP method for assigning precursor multiplicity to further examples seems desirable, however, in the light of the conflicting conclusions reached on the basis of photosensitization and quenching studies (see, for example, Bartlett and Engel, 1968; Abram et al., 1969).

Thermolysis of cis-4 in the n.m.r. probe leads to a spectrum in which the methyl groups of trans-4 manifest themselves in emission (Porter et al., 1972). This is clear evidence of a stepwise cleavage of nitrogen from the azo-compound. Application of Kaptein's rules, indicates that g_{PhCMe_2}. > g_{PhNN}.. The failure to observe CIDNP evidence for intermediate diazenyl radicals in the thermolysis of 5 may indicate a very short lifetime in that system or, in the limit, a concerted cleavage of both carbon–nitrogen bonds. This is quite reasonable in view of the stability of both diphenylmethyl and benzyl radicals compared to phenyl radicals. Thermal decomposition of diazoaminobenzene (PhN:N.NHPh) enriched in ^{13}C and ^{15}N leads to polarization of both nuclei in the reactant, and this is claimed to be a more sensitive procedure than ^1H CIDNP (Lippmaa et al., 1971). CIDNP thus provides a convenient alternative to following the racemization of optically active azo-compounds (Tsolis et al., 1972) for studying the behaviour of diazenyl radicals.

Diazenyl radicals have also been detected in related systems. The rapid rearrangement of 1,3,5-triarylpentazadienes [equation (47)] involves intermediate triazenyl-diazenyl radical pairs, as indicated by the appearance in emission of the n.m.r. transitions of the p-methyl protons of the starting material when $Ar^1 = Ar^3 = p\text{-}CH_3.C_6H_4$ (Hollaender and Neumann, 1970). The weak emission of benzene which accompanies a much more intense emission due to toluene when the 1,3-diaryltetrazene 6 decomposes in acetone at 50° has been interpreted

(Hollaender and Neumann, 1971) as supporting a free-radical induced decomposition of intermediate phenyldi-imide (phenyldiazene)[8] by way of the diazenyl radical [equation (48); cf. Hoffmann and Guhn, 1967].

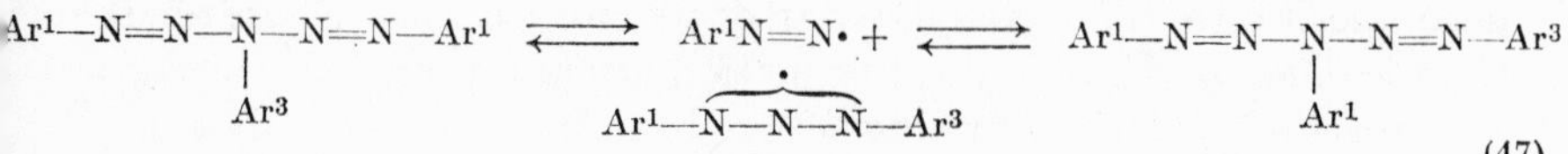

$$Ar^1-N{=}N-N-N{=}N-Ar^1 \;\rightleftharpoons\; Ar^1N{=}N{\cdot} + \;\rightleftharpoons\; Ar^1-N{=}N-N-N{=}N-Ar^3$$

$$\underset{\textstyle Ar^3}{\textstyle |}\qquad\qquad\qquad \overset{\textstyle \cdot}{\overline{Ar^1-N-N-N-Ar^3}}\qquad\qquad \underset{\textstyle Ar^1}{\textstyle |}$$

(6) $\hspace{8.5cm}$ (47)

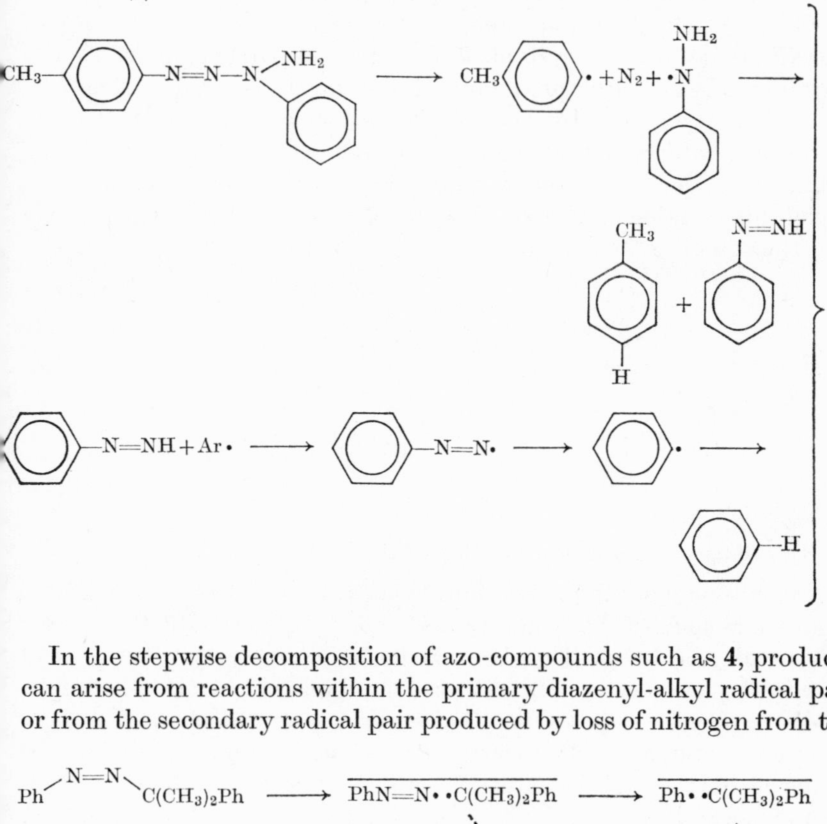

(48)

In the stepwise decomposition of azo-compounds such as **4**, products can arise from reactions within the primary diazenyl-alkyl radical pair or from the secondary radical pair produced by loss of nitrogen from the

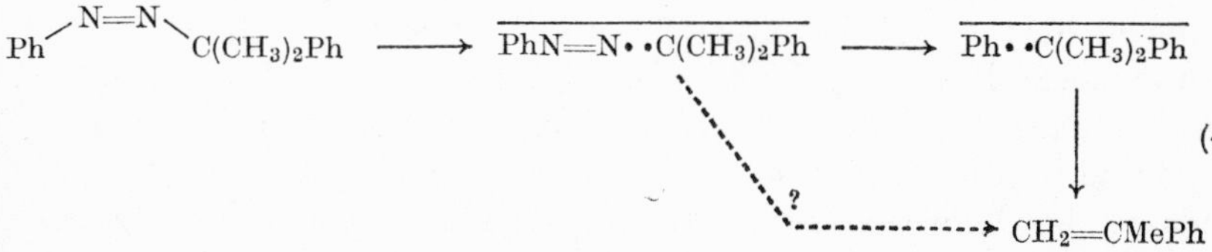

(49)

diazenyl moiety. Thus, for example, α-methylstyrene could be produced from cis-**4** by way of either or both types of radical pair [equation (49)].

[8] Polarization of benzene (E) produced in the thermal decomposition of the phenyldi-imide derivative $Ph.N{=}N.CO_2^-K^+$ has also been reported (Heesing and Kaiser, 1970).

The observation (Porter *et al.*, 1972) that added $BrCCl_3$ almost completely suppresses the polarization of the olefin, while leaving the polarization of trans-**4** unaffected, points to the secondary radical pair as the principal immediate precursor of α-methylstyrene. A rate constant for the decomposition of the diazenyl radical of 10^7–$10^9\,sec^{-1}$ has been estimated. Cage collapse and free-radical formation are also thought to occur and appropriately polarized products have been identified (see above).

Cyclic azo-compounds are potential sources of biradicals, the chemistry of which is a subject of continuing interest. To date the only example of CIDNP studies on such systems is the report of strong emission signals corresponding to dimeric products of the biradical **7** (related to trimethylenemethane) generated by thermolysis of **8** (Berson *et al.*, 1971). The observation accords with theory for $T_{\pm 1}$–S mixing (Closs, 1971a, c; Kaptein *et al.*, 1971), but it does not distinguish with certainty a concerted triplet dimerization from a stepwise reaction of singlet and triplet via a further biradical intermediate.

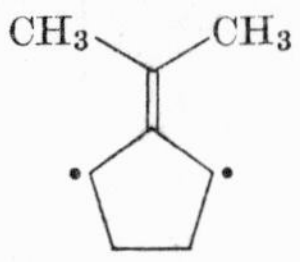

(7) 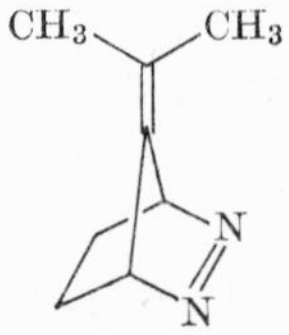   (8)

2. *Arenediazonium and related compounds*

Arenediazonium compounds and arylacylnitrosamines have long been regarded as sources of free aryl radicals (for reviews, see Rüchardt *et al.*, 1970; Cadogan, 1970). Thus benzenediazonium ion, when treated with base in the presence of benzene, produces biphenyl by attack of intermediate phenyl radicals on the aromatic hydrocarbon (Gomberg reaction). Similarly, phenylation of aromatic compounds can be achieved by thermal decomposition of N-nitrosoacetanilide (**9**). This compound undergoes a preliminary rearrangement to the diazoacetate (**10**) (which is in equilibrium with an ionic form) prior to the formation of phenyl radicals [equation (50)]. Aromatic diazonium groups can be replaced by hydrogen by a variety of reducing agents (Meerwein

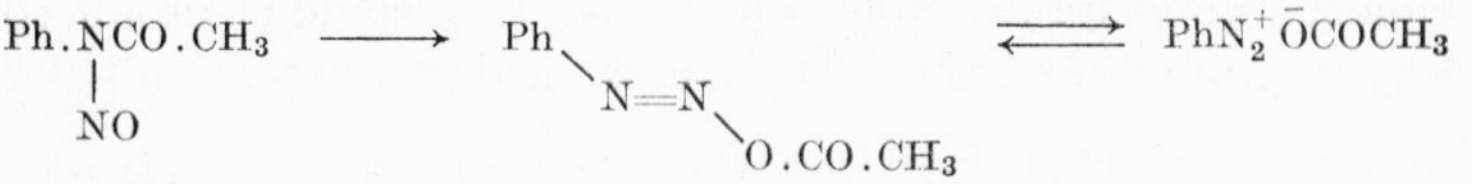

(50)

(9)   (10)

reduction) and, in neutral or mildly basic conditions, radicals are thought to be involved. Formation of aryl radicals from arenediazonium ions is thought to involve the reaction sequence shown in Scheme 6.

A continuing problem in this area has been the identification of the species X of Scheme 6, which plays a key role in the chain mechanism. The favoured view is that X^- is the diazotate ion, $ArN{=}NO^-$.

Benzene produced by treatment of benzenediazonium salts with base (e.g., NaOH/CH$_3$CN) (Lane *et al.*, 1969; Rieker *et al.*, 1969, 1971; Rieker, 1971; Levit *et al.*, 1971, 1972; Bilevich *et al.*, 1970, 1971) or by

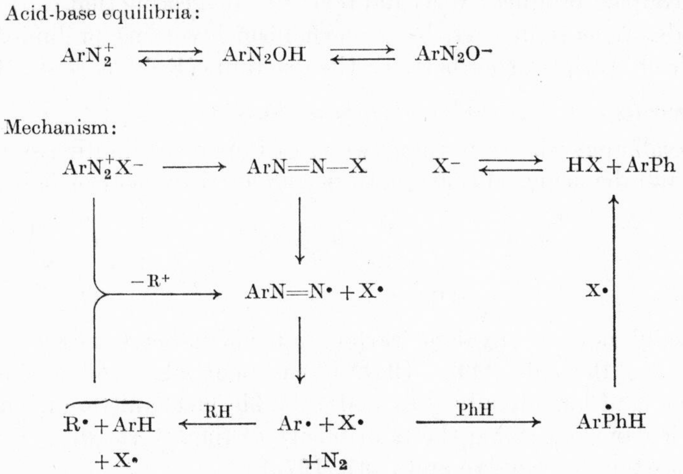

Acid-base equilibria:

$$ArN_2^+ \;\;\rightleftharpoons\;\; ArN_2OH \;\;\rightleftharpoons\;\; ArN_2O^-$$

Mechanism:

SCHEME 6.

thermal decomposition of N-nitrosoacetanilide (Kasukhin *et al.*, 1970; see also, Koenig and Mabey, 1970) shows pronounced transient emission in its ^{1}H n.m.r. spectrum. Analogous observations have been reported in the reaction of diphenyliodonium iodide (Ph$_2$I$^+$I$^-$) with bases such as sodium ethoxide or tertiary amines (Bilevich *et al.*, 1971). Polarization (E) is also observed in the ^{13}C n.m.r. spectrum of anisole produced on treatment of 4-methoxybenzenediazonium tetrafluoroborate with sodium hydroxide, whereas enhancement is seen in the ^{13}C resonance of C-1 in the reactant ion (Berger *et al.*, 1972). Polarization of ^{13}C and ^{15}N nuclei has been reported in the reaction of an isotopically enriched arenediazonium salt with sodium phenoxide (Lippmaa *et al.*, 1971).

While authors have generally been cautious in drawing mechanistic conclusions from their observations, application of Kaptein's rules to the ^{1}H n.m.r. results indicates that Δg must be negative in the radical-pair precursor of benzene. Relevant g-values are as follows: PhNN$\cdot \leqslant 2{\cdot}0012$; Ph$\cdot$, $2{\cdot}0020$; PhN{=}NO$\cdot$, $2{\cdot}0017$; HO$\cdot$, $2{\cdot}0118$. Since the ^{13}C n.m.r.

results suggest that nitrogen is retained in the radical pair, either HO·
or Ph.N=NO·as the gegen radical would lead to $\Delta g < 0$, although in
the latter case the value would be very close to zero. On the other hand,
products of free-radical hydroxylation are not observed. Further
carefully chosen CIDNP studies could clearly lead to an unequivocal
identification of X.

Arenediazonium ions can, of course, bring about electrophilic aromatic substitution giving aromatic azo-compounds. Using $Ph\overset{+}{N}\equiv N$ and
PhO^-, polarized signals have been observed in the ^{15}N-spectrum (6 MHz)
of the coupled product (A, A) and reactant, suggesting that the reaction
proceeds, at least in part, by a mechanism involving preliminary reversible electron transfer between the reactants (Bubnov et al., 1972).

3. Diazoalkanes, diazirines and carbene reactions

Diazoalkanes (shown in their two most important limiting structures
in 11) and diazirines (12) are both decomposed by heat or light giving

$$\begin{array}{ccc} R^1 \\ \diagdown \\ {}^+{}^- \\ C=N=\overset{..}{N} \\ \diagup \\ R^2 \end{array} \longleftrightarrow \begin{array}{c} R^1 \\ \diagdown \\ \overset{-}{C}-\overset{+}{N}\equiv N \\ \diagup \\ R^2 \end{array} \qquad \begin{array}{c} R^1 \\ \diagdown \quad N \\ C \parallel \\ \diagup \quad N \\ R^2 \end{array}$$

$$(11) \qquad\qquad\qquad (12)$$

electrically neutral divalent carbon intermediates (carbenes) (for a
review, see Bethell, 1969). CIDNP has been observed during such
reactions and has already provided valuable mechanistic information,
in particular, concerning the multiplicity of the intermediate carbenes
(see, for example, Kaplan and Roth, 1972).

The central carbon atom of carbenes, by virtue of its divalency,
possesses two non-bonding electrons and has available two low-lying
orbitals in which they can be accommodated. Thus both singlet and
triplet electronic states are accessible, the triplet state being usually of
lower energy except in the case of dihalogenomethylenes. An important
problem in carbene chemistry is to decide which of the two states is
involved in a particular chemical process. In general, the singlet state
is thought to be capable of coordination with non-bonding electron
pairs, direct insertion into covalent bonds [equation (51)] and stereospecific addition to olefins. The triplet states of carbenes are generally
thought to show radical-like properties, inserting into covalent bonds
by atom abstraction followed by radical pair recombination [equation
(52)];[9] addition to olefins is stepwise and involves an intermediate 1,3-
biradical. Interpretations are complicated by the possibility of interconversion of spin states of carbenes at rates comparable to those of
their reactions with other molecules.

[9] Note that equation (52) implies T–S mixing prior to radical recombination.

$$R^1R^2C \colon^S + HC\diagdown \longrightarrow \left[\begin{array}{c} H\text{-----}C\diagdown \\ \diagdown \quad \diagup \\ R^1R^2C \end{array} \right] \longrightarrow HCR^1R^2 \cdot C\diagdown \qquad (51)$$

$$R^1R^2C \colon^T + HC\diagdown \longrightarrow R^1R^2CH \cdot \cdot C\diagdown \longrightarrow HCR^1R^2 \cdot C\diagdown \qquad (52)$$

Irradiation of a solution of diazomethane $(11; R^1 = R^2 = H)$ in toluene containing a triplet sensitizer, benzophenone, leads to the production of ethylbenzene by insertion of methylene into the benzylic C—H bond. The ^1H n.m.r. signal of the terminal methyl group is strongly polarized A/E, consistent with hydrogen abstraction by triplet methylene followed by collapse of the resultant radical pair $(\Gamma_{ME} = + + - - - + - = - \equiv$ A/E) (Roth, 1972b). In contrast, no polarization is observed in the absence of a sensitizer, despite the fact that ethylbenzene is produced. This is consistent with a direct insertion of singlet methylene into the C—H bond, but it could also arise from an abstraction-recombination mechanism if the lifetime of the intermediate radical-pair were too short to permit a significant amount of T_0–S mixing.

With deuteriochloroform, singlet and triplet methylene, produced by photolysis of diazirine $(12; R^1 = R^2 = H)$, lead to different polarized products (Roth, 1971b). Sensitization by benzophenone gives the C—D insertion product $CH_2D . CCl_3$ in which the protons appear as a triplet, as a result of coupling with deuterium, showing enhanced absorption $(\Gamma_{NE} = + + - - - + \equiv A)$. Polarized $CH_2D . CCl_3$ is not observed in the absence of sensitizer; instead, the C—Cl insertion product $CH_2Cl . CDCl_2$ appears in emission $(\Gamma_{NE} = - + - - - = - \equiv E)$. Clearly, an abstraction-recombination sequence is necessary to account for the polarization. Such mechanisms for insertion of carbenes into carbon–halogen bonds have previously been proposed on the basis of kinetic and product evidence (Bamford and Casson, 1969; Clark et al., 1970). A similar pattern of behaviour has been described in the reaction of methylene (singlet and triplet) with perdeuterio-2-chlorobutane. In addition to insertion products formed by collapse of the primary radical pair, disproportionation products (e.g., $\overline{ClH_2C \cdot \cdot CD(CD_3)CD_2CD_3}^S \rightarrow$ $CH_2DCl + CD_3CD{=}CDCD_3$; $\overline{H_2DC \cdot \cdot CCl(CD_3)CD_2CD_3}^S \rightarrow CH_2D_2 +$ $CD_3CCl{=}CD . CD_3$) were also formed, their origin being revealed by the phase of the ^1H polarization. Polarization in radicals escaping the primary cage was in some cases retained in the products of the decomposition of further diazoalkane which the free radicals induced (Roth, 1972a).

The reaction of photochemically generated methylene with carbon tetrachloride giving pentaerythrityl tetrachloride (13) in high yield has long been recognized as a chain reaction on the basis of its high quantum yield for consumption of diazomethane (Urry and Eiszner, 1952; Urry *et al.*, 1957). The reaction is plausibly interpreted as being initiated by abstraction of chlorine from CCl_4 by methylene. The chain nature of reaction of CCl_4 (and $CHCl_3$) with methoxycarbonylmethylene produced photochemically from 11 ($R^1 = H$; $R^2 = CO.OCH_3$) has recently been demonstrated by CIDNP. However, when irradiation is interrupted, the decay of the polarization of the C—Cl insertion products is much slower than expected on the basis of spin-lattice relaxation times (Cocivera and Roth, 1970). By examination of the effect of triplet photosensitizers on the polarization of 14 in the direct photolysis of diazomethane in carbon tetrachloride solution, it has been established that the reaction is initiated by chlorine abstraction by *singlet* methylene, (polarization, E; with benzophenone, A) (Roth, 1971a, c). Most of the 14 observed arises by recombination of the initial radical pair ($\Gamma_{NE} = - + - - = - \equiv E$). Escape of $Cl_3C\cdot$ radicals, however, permits them to react with diazomethane, as shown in Scheme 7. It is noteworthy that 13 and the intermediate products are also polarized: 13,

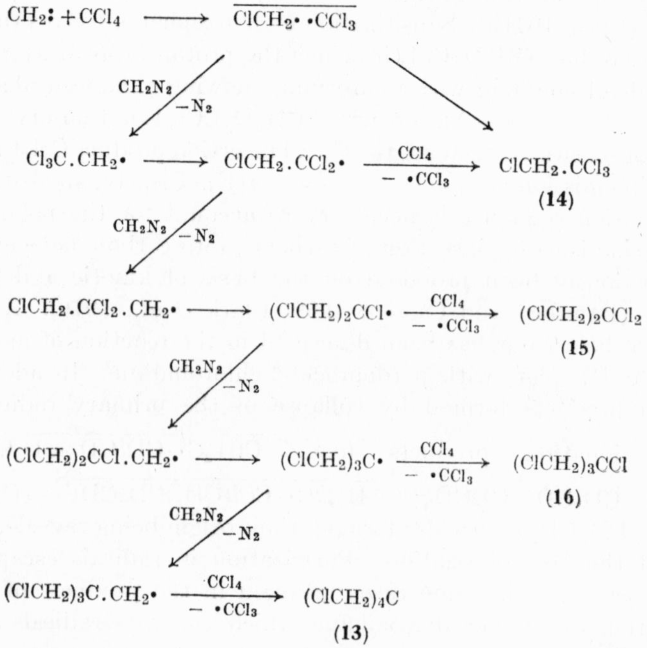

$$CH_2\colon + CCl_4 \longrightarrow \overline{ClCH_2\cdot\ \cdot CCl_3}$$

$$Cl_3C.CH_2\cdot \longrightarrow ClCH_2.CCl_2\cdot \xrightarrow[-\ \cdot CCl_3]{CCl_4} ClCH_2.CCl_3$$
$$(14)$$

$$ClCH_2.CCl_2.CH_2\cdot \longrightarrow (ClCH_2)_2CCl\cdot \xrightarrow[-\ \cdot CCl_3]{CCl_4} (ClCH_2)_2CCl_2$$
$$(15)$$

$$(ClCH_2)_2CCl.CH_2\cdot \longrightarrow (ClCH_2)_3C\cdot \xrightarrow[-\ \cdot CCl_3]{CCl_4} (ClCH_2)_3CCl$$
$$(16)$$

$$(ClCH_2)_3C.CH_2\cdot \xrightarrow[-\ \cdot CCl_3]{CCl_4} (ClCH_2)_4C$$
$$(13)$$

(with $CH_2N_2 / -N_2$ steps connecting the stages)

SCHEME 7.

weak A; 15, weak A; **16**, very weak E. Since, on the basis of Scheme 7, all are *e*-products of the primary radical pair, they should show the opposite polarization to **14**, with low intensity due to the relaxation of nuclear spin states which occurs concurrently with reaction in free radicals. Triplet photosensitization converts the polarization of **13** to E, but the polarization of **15** and **16** is unchanged. The tentative and not wholly satisfactory conclusion is that **13** is polarized in the reaction of **16** with further methylene, but that polarized **15** and **16** arise from random recombination of free radicals (e.g., $ClCH_2 \cdot + \cdot CCl_2 . CH_2Cl \rightarrow$ **15**).

Insertion reactions of diphenylmethylene have also been studied using the CIDNP technique. Kinetic and product studies on reactions of diphenylmethylene have suggested that reversible interconversion of singlet and (ground) triplet states is rapid compared to carbene-consuming reactions (Bethell *et al.*, 1965, 1970; Closs, 1968), but observed polarizations usually arise from reactions of the triplet. Insertion of thermally or photochemically generated diphenylmethylene into the benzylic C—H bonds of toluene gives 1,1,2-triphenylethane with the multiplet due to the aliphatic A_2B system of protons polarized A/E (Closs and Closs, 1969a). This is the reverse of the polarization of the same product generated from $Ph_2CH . N : N . CH_2Ph$ where the primary radical pair is in a singlet state (see Section V,C,1). The changes in polarization resulting from alteration in the *g*-values of the radicals in the primary pair by introduction of substituents into the carbene have been reported for both ^1H- (Closs and Trifunac, 1970b, c) and ^{19}F-n.m.r. spectra (Brinkman *et al.*, 1973b; Bethell *et al.*, 1972b). Because of the high sensitivity of the chemical shift of the ^{19}F nucleus to its chemical environment, ^{19}F n.m.r. spectroscopy has considerable potential in mechanistic studies. For example, insertion of triplet diarylmethylenes into the benzylic C—H bonds of benzyl fluoride produces $ArAr'CH . CHFPh$ having two chiral centres and hence *erythro*- and *threo*-diastereoisomers. The quartets due to the fluorine part of the AMX multiplets of the diastereoisomers are clearly resolved, and are formed in equal amounts polarized identically AAEA when $Ar = C_6H_5$, $Ar' = C_6H_4Cl$. *o*-Fluorosubstituents in the carbene result in an additional doublet splitting of each line which is only detectable in the polarized spectrum.

Analogous to the decomposition of diazoalkanes, thermolysis of azides leads to the production of nitrenes. This is exemplified for ethyl

$$EtO.CO.N_3 \longrightarrow EtO.CO.\ddot{N}: \longrightarrow \quad\quad\quad\quad\quad\quad\quad (53)$$

NH.CO.OEt

azidoformate in equation (53). Insertion of the nitrene into the tertiary C—H bond of *trans*-decalin gives rise to polarization of the signal due to the proton attached to nitrogen, its phase indicating reaction via the triplet ($\Gamma_{NE} = + + + - = - \equiv E$) (Brinkman *et al.*, 1973a).

D. *The Photolysis of Aldehydes and Ketones*

1. *Introduction*

The photolysis of carbonyl compounds is one of the most intensively studied areas of photochemistry. Since CIDNP studies have been concerned mostly with aldehydes and ketones we shall confine these brief introductory remarks to such compounds. More extensive reviews are available (e.g., Simons, 1971).

The long-wavelength absorption band of aldehydes and ketones corresponds to a forbidden n–π^* transition, typically in the range 280–340 nm. The allowed π–π^* transition occurs at shorter wavelengths. Intersystem crossing from the first excited singlet state to the lowest triplet state,[3] (n–π^*), is usually very efficient, and several of the characteristic photochemical reactions of carbonyl compounds involve this excited species. Competing with chemical reaction is triplet energy transfer.

Typical chemical reactions of photoexcited aldehydes and ketones are cleavage reactions, usually designated as Norrish Type I [equation (54)], II [equation (55)] and III [equation (56)], hydrogen abstraction [equation (57)] and cycloadditions, such as the Paterno-Büchi reaction [equation (58)]. Of these, Norrish Type II cleavage and the related

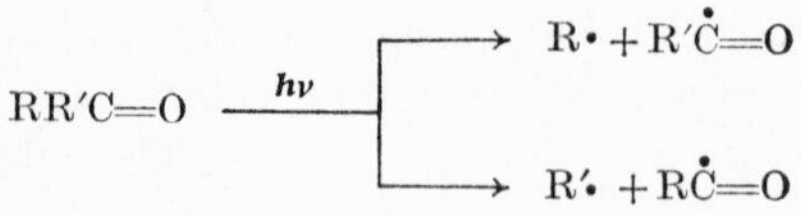

$$\text{RR'C=O} \quad \xrightarrow{h\nu} \quad \begin{cases} \longrightarrow \text{R} \cdot + \text{R'}\overset{\bullet}{\text{C}}\text{=O} \\ \\ \longrightarrow \text{R'} \cdot + \text{R}\overset{\bullet}{\text{C}}\text{=O} \end{cases} \tag{54}$$

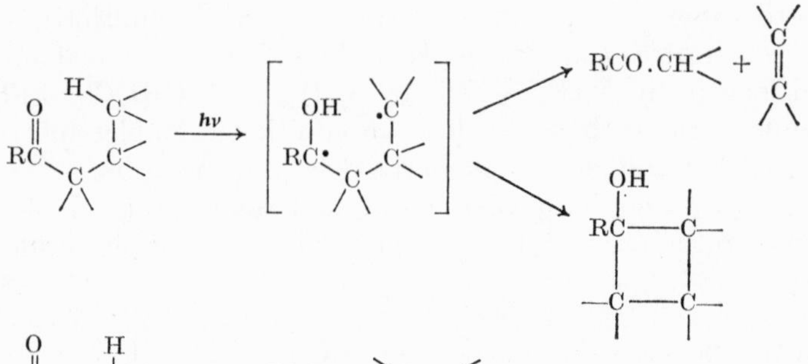

(55)

$$\tag{56}$$

$$RR'CO + R''H \xrightarrow{h\nu} RR'\dot{C}-OH + R''\cdot \longrightarrow RR'R''COH, [RR'C(OH)]_2, R''R'' \text{ etc.}$$
(57)

$$RR'CO + {>}C{=}C{<} \xrightarrow{h\nu} \begin{array}{c} | \quad | \\ -C-C- \\ | \quad | \\ RR'C-O \end{array}$$
(58)

cyclization and oxetane formation are thought to involve 1,4-biradical intermediates. CIDNP has not been reported in such reactions and this may be a consequence of the inability of the radical centres to diffuse far enough apart for T_0–S mixing to take place effectively. T_{-1}–S mixing could lead to polarization in such systems (Closs, 1971a, c; Kaptein et al., 1971). Most CIDNP studies have been concerned with Norrish Type I cleavage and abstraction from hydrogen donors.

2. Norrish Type I cleavage

Norrish Type I cleavage is the principal photochemical reaction of aliphatic ketones. Predictably, the reaction is important where cleavage leads to relatively stable t-alkyl or benzylic radicals and where the carbonyl compound possesses no γ-hydrogen atom. The direction of cleavage, as well as the relative importance of competing processes, seems to be wavelength-dependent. Pinacolone, $CH_3.CO.C(CH_3)_3$, serves as a simple example since photolysis leads to Type I cleavage only. From a study of the non-linear effect of varying triplet quencher (piperylene) concentrations on the quantum yield for pinacolone photolysis at 313 nm, it has proved possible to demonstrate that cleavage occurs in both $^1(n{-}\pi^*)$ and $^3(n{-}\pi^*)$ excited states (Yang and Feit, 1968; Yang et al., 1970). CIDNP is observed on photolysis (Blank et al., 1971) in the products, acetaldehyde, isobutene, isobutane, biacetyl and hexamethylethane, and in the reactant ketone. The polarizations can be interpreted in terms of Scheme 8 using known values of g and a_H for acetyl and t-butyl radicals, provided that products are regarded as arising from triplet correlated radical pairs or from uncorrelated pairs. Free radical scavengers (e.g., tri-n-butyltin hydride) and triplet quenchers (e.g., piperylene) remove biacetyl and hexamethylethane from the products and reduce the intensities of the other polarizations but not their phases, even at high concentration. Thus CIDNP gives no evidence for singlet precursors in the cleavage reaction. The reason for this discrepancy between CIDNP and conventional photochemical techniques needs further study.

The CIDNP technique has led to similar conclusions about the precursor multiplicity for methyl benzyl ketone (Blank et al., 1971). With dibenzyl ketone and phenyl α-phenylethyl ketone (Closs, 1971a; Müller

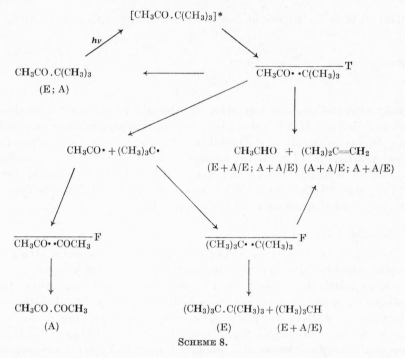

SCHEME 8.

and Closs, 1972), the primary radical pair, again believed to be triplet, contains benzyl and phenylacetyl radicals, the latter losing carbon monoxide in 10^{-8} sec to give a secondary radical pair. Bibenzyl, a major product of photolysis, cannot, however, arise from this since it appears polarized A in the ^1H n.m.r. spectrum, an impossibility for a precursor pair with $\Delta g = 0$, unless a "memory effect" operates. A different conclusion has been reached in the photolysis of di-t-butyl ketone, where the disproportionation products isobutane and isobutene are thought to be formed from an encounter pair of t-butyl radicals ($\Delta g = 0$; perfect A/E multiplet polarization), with some isobutene arising from disproportionation in the primary pivaloyl-t-butyl radical pair ($\Delta g > 0$; enhanced absorption) (Tomkiewicz et al., 1971, 1972).

With both benzyl ketones, the parent ketone shows very strong emission in the ^1H n.m.r. spectrum during photolysis (Blank et al., 1971; Closs, 1971a; Closs and Paulson, 1970) and this may mean that cleavage/recombination is normally an important route for the decay of excited triplet states of such ketones to their ground states (see also Rosenfeld et al., 1972). However, in the case of substituted benzoin methyl ethers, ArCO.CH(OMe)Ar', the recombination of *free* ArCO· and ·CH(OMeAr) radicals seems to be responsible for the polarization of the reactant since cross-products are detected when a mixture of two ketones is irradiated (DoMinh, 1971).

A singlet precursor has, however, been proposed for 4,4,4-trichloro-2-methyl-1-butene (17) produced in the photolysis of isomesityl oxide (18) in carbon tetrachloride solution on the basis of its all-emission spectrum (DoMinh, 1971). There remains some ambiguity, however, about the detailed route by which 17 is formed. Moreover there is other evidence suggesting that ketone photolysis in carbon tetrachloride is different

$$Cl_3C.CH_2C(Me):CH_2 \qquad\qquad CH_3.CO.CH_2.C(Me):CH_2$$
$$(17) \qquad\qquad\qquad (18)$$

from photolyses in other solvents (den Hollander, 1971; Kaptein, 1971b; den Hollander et al., 1971). Thus, photolysis of di-isopropyl ketone in, for example, deuteriochloroform, benzene or cyclohexane gives polarized isobutyraldehyde, chloroform, mono-deuteriomethylene chloride, propene, propane and/or 2,3-dimethylbutane. The polarization of all these products is consistent with a primary isopropyl-isobutyryl radical pair which is triplet. However, in carbon tetrachloride as solvent, the products chloroform, propene, isopropyl chloride and trichloroisobutane all show polarization indicating a singlet precursor: only a weakly enhanced absorption of the aldehydic proton of isobutyraldehyde indicates any reaction by way of a triplet precursor. Significantly, the starting ketone is not polarized in carbon tetrachloride although its methyl groups show enhanced absorption in CDCl$_3$. It has been suggested that the effect of carbon tetrachloride is to form an exciplex with the first excited singlet state of the ketone and that the singlet precursor, $\overline{(CH_3)_2CH\cdot\cdot CCl_3}^S$, is formed directly from this, together with isobutyryl chloride. In support of this interpretation, carbon tetrachloride is observed to quench the fluorescence of di-isopropyl ketone (see also Tomkiewicz et al., 1972). In the photolysis of di-t-butyl ketone in C$_7$F$_{14}$, the relatively small effect of carbon tetrachloride (0·5 M) has been attributed to trapping of free-radicals. The supporting observation that carbon tetrachloride suppresses the formation of t-butyl radicals, as detected by e.s.r. experiments using a flow system, is not inconsistent, however, with singlet exciplex formation which could well reduce the number of radicals escaping from the primary pairs to below the level of detection.

Norrish Type I cleavage of cyclic ketones necessarily yields biradicals, and in certain cases (e.g., cycloheptanone, camphor) strong emissions due to T$_{-1}$–S mixing have been reported (Closs and Doubleday, 1972).

3. Hydrogen abstraction (photoreduction)

Excitation of the n–π^* transition of carbonyl compounds entails transfer of an electron from an orbital essentially localized on oxygen to an orbital delocalized over the π-system. In aliphatic carbonyl compounds this π-system is merely that of the carbonyl group, but in conjugated compounds more extensive delocalization becomes possible. The result of the electronic transition is thus to produce an unpaired electron on oxygen so that n–π^* excited states, especially the triplet, can behave chemically rather like alkoxy-radicals. In particular, hydrogen atom abstraction becomes possible (cf. Norrish Type II cleavage) and this kind of process has provided a number of examples of CIDNP.

Thus irradiation of benzophenone in toluene gives the photoreduction product **19**, benzpinacol, and bibenzyl [equation (59)] of which only **19** is polarized A $(\Gamma_{NE} = + + + - - = + \equiv A)$. Likewise, the analogous product from ethylbenzene in place of toluene contains an A_3B system of protons with the methine quartet polarized A/E $(\Gamma_{ME} = + + - + + +$ $= - \equiv A/E)$ (Closs and Closs, 1969b). The patterns of polarizations show the expected changes when substituents such as halogens are introduced into the aromatic rings, so modifying Δg (Closs *et al.*, 1970). Indeed, by computer simulation it is possible to derive isotropic g-values for substituted diphenylmethyl radicals in good agreement with values obtained directly from e.s.r. measurements (Closs, 1971a).

$$Ph_2CO + PhCH_3 \xrightarrow{h\nu} \overline{Ph_2C(OH)\cdot\,\cdot CH_2Ph}^{\,T} \longrightarrow$$

$$Ph_2C(OH)CH_2Ph + PhCH_2CH_2Ph + Ph_2C(OH)C(OH)Ph_2 \quad (59)$$
$$(19)$$

The photolysis of benzaldehydes in which a second aldehyde molecule acts as hydrogen donor shows slightly different CIDNP behaviour (Closs and Paulson, 1970; Cocivera and Trozzolo, 1970). The aldehydic proton of the starting material appears in the ^1H n.m.r. spectrum with strongly enhanced absorption, while other polarized signals arise from the product benzoins [equation (60)]. Piperylene quenches the polarization, suggesting a triplet precursor. Changing the solvent from cyclohexane to p-bromotoluene has no effect on the pattern of polarization, indicating that abstraction of hydrogen from the solvent, which would involve a change in g for the solvent radicals, plays no part. Thus the primary radical pair is presumably $\overline{ArCH(OH)\cdot\cdot COPh}^{\,T}$. Polarized ArCHO could clearly arise by reversal of the process of formation of this radical pair, but application of Kaptein's rule gives $\Gamma_{NE} = + + + - =$ $- \equiv E$. It is concluded that polarized aldehyde arises by hydrogen atom transfer from a free hydroxybenzyl radical to a further molecule of the aldehyde. Confirmation of this comes from the observation that the relative intensities of the aldehydic and *ortho* ring protons in the product are quantitatively related to the free radical lifetime which is a function of the concentration of the aldehyde.

$$2ArCHO \xrightarrow{hv} \overline{ArCH(OH) \cdot \cdot COAr}^T \longrightarrow ArCH(OH)COAr \qquad (60)$$

The excited triplet states of quinones can be fairly readily populated by irradiation and nuclear polarization observed (Cocivera, 1968). Hydrogen atom abstraction leads to the relatively stable semiquinone radicals and, in alkaline media, radical anions. Recombination of radical pairs formed in this way can give rise to CIDNP signals, as found on irradiation of phenanthraquinone (20) in the presence of donors such as fluorene, xanthene and diphenylmethane (Maruyama *et al.*, 1971a, c; Shindo *et al.*, 1971; see also Maruyama *et al.*, 1972). The adducts are believed to have the 1,2-structure (21) with the methine proton appearing in absorption in the polarized spectrum, as expected for a triplet precursor. Consistently, *thermal* decomposition of 21 as shown in equation (61) leads to polarization of the reactant but now in emission (Maruyama

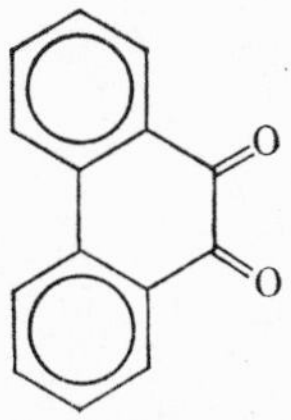

(20)

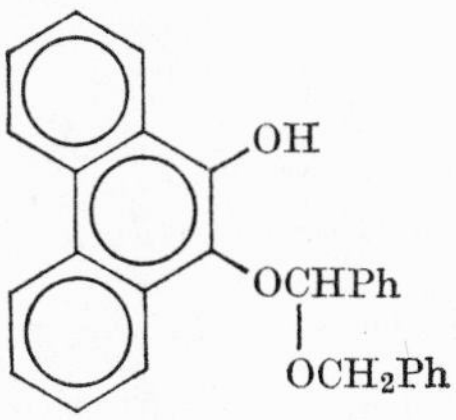

(22)

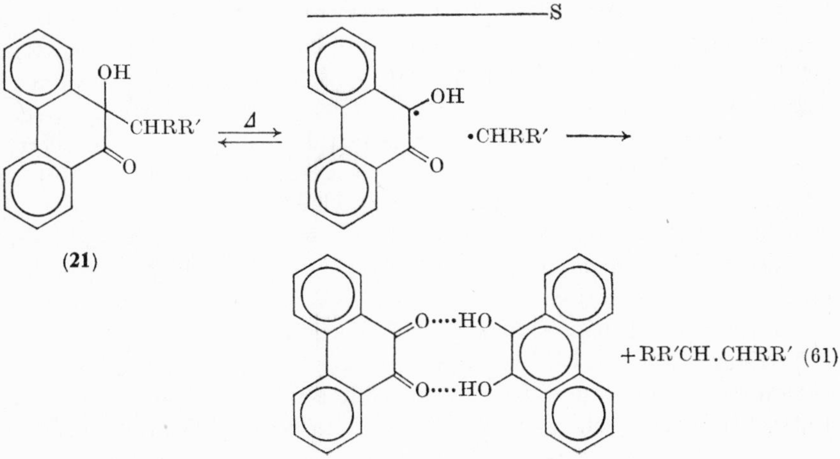

(21)

+RR'CH.CHRR' (61)

et al., 1971b). Photo-reduction of phenanthraquinone with dibenzyl ether gives the 1,4-adduct **22** with the methine proton polarized E (Maruyama and Otsuki, 1971).

One potentially important example of CIDNP in products resulting from a radical pair formed by electron transfer involves a quinone, anthraquinone β-sulphonic acid (**23**). When irradiated in the presence of the *cis-syn* dimer of 1,3-dimethylthymine (**24**), enhanced absorption due to vinylic protons and emission from the allylic methyls of the monomer (**25**) produced can be observed (Roth and Lamola, 1972). The phase of the polarizations fits Kaptein's rules for intermediate formation of $\overline{23 \cdot 24 \dagger}^T$ produced by electron abstraction by the first excited triplet state of the quinone. The result is believed to be relevant to the photoenzymatic reversal of pyrimidine dimer formation in deoxyribonucleic acid.

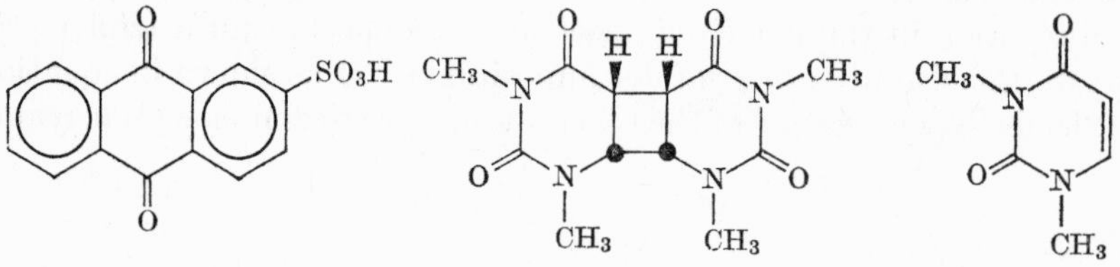

(23)   (24)   (25)

E. *Reactions Involving Organometallic Compounds*

1. *Reaction of lithium alkyls with organic halides*

The first example of chemically induced multiplet polarization was observed on treatment of a solution of n-butyl bromide and n-butyl lithium in hexane with a little ether to initiate reaction by depolymerizing the organometallic compound (Ward and Lawler, 1967). Polarization (E/A) of the protons on carbon atoms 1 and 2 in the 1-butene produced was observed and taken as evidence of the correctness of an earlier suggestion (Bryce-Smith, 1956) that radical intermediates are involved in this elimination. Similar observations were made in the reaction of t-butyl lithium with n-butyl bromide when both 1-butene and isobutene were found to be polarized. The observations were particularly significant because multiplet polarization could not be explained by the electron–nuclear cross-relaxation theory of CIDNP then being advanced to explain net polarization (Lawler, 1967; Bargon and Fischer, 1967).

Tertiary amines also depolymerize lithium alkyl tetramers and hexamers and can be used to trigger reactions with alkyl halides (Lepley, 1968). However, when triethylamine is used to initiate the butyl

bromide–butyl lithium reaction, some N,N-diethylvinylamine, $Et_2N.$
$CH{=}CH_2$, polarized A/E, is formed. The phase of this polarization is
consistent with polarization in an encounter pair, $\overline{Et_2NCH(CH_3){\cdot}{\cdot}C_4H_9}^F$,
an explanation which is preferable to the earlier suggestion of polariza-
tion in radical transfer reactions. The reverse phase in the olefinic
products thus corresponds to disproportionation within a geminate
pair of butyl radicals.

Lithium alkyls also bring about polarization during reaction with
other halides. Thus, in the reaction of n-butyl lithium with s-butyl
iodide, polarization of the α-protons of both s-butyl iodide (E/A) and
n-butyl iodide (A/E) occurs with comparable intensity (Ward et al.,
1969b; Lepley, 1969c). Clearly, metal–halogen interchange occurs
under the reaction conditions. By carrying out a related reaction (t-
butyl lithium + n-butyl iodide) in ether at $-70°$, when metal–halogen
interchange occurs without disproportionation or combination of alkyl
groups, it has proved possible to establish that this process does not
lead to polarization (Ward, 1971). Thus polarization occurs in the
course of the coupling and elimination reactions of the alkyl moieties,
including the reaction leading to re-formation of the alkyl iodide.
Interestingly, no polarization of the lithium alkyl is observed (Ward
et al., 1969b; Lepley and Landau, 1969).

Polarization is found in reactions involving chlorides. 1,1-Dichloro-
2,2-dimethylcyclopropane (26) reacts with lithium ethyl in benzene–
ether solution (40°) giving mainly 1-chloro-2,2-dimethylcyclopropane
(27; X = H) and 3-methyl-1,2-butadiene (28) both of which are polarized
(Ward et al., 1968). If n- or t-butyl lithium are used in the reaction, the
butene produced by disproportionation shows only net polarization,

(26)   (27)   (28)

indicating that g-effects dominate in the precursor radical pair. Clearly,
this observation together with the production of (27; X = H) points to
an α-chlorocyclopropyl radical intermediate. The route by which this
radical is converted into 28 is not altogether clear, but it could involve
(27; X = Li). α-Halolithium compounds ("carbenoids") can be trapped
using carbon dioxide in related reactions. They could be formed from
the geminate radical pair in which nuclear spin selection occurs, with
polarization being retained during the subsequent rapid α-elimination
and rearrangement. More detailed investigation is desirable, however,

since polarization of lithium-containing species has not been observed in this or other systems (see above).

Because of the polymeric nature of lithium alkyls in hydrocarbon solvents, it is conceivable that the radical species produced are in some way associated with the organometallic aggregates. This seems unlikely in the light of observations that the pattern of polarization in, for example, the products of the reaction of ethyl lithium and isopropyl iodide (benzene solution; 40°C) is virtually identical with that in the products from thermolysis of propionyl peroxide in the presence of isopropyl iodide (o-dichlorobenzene solution; ca. 120°C) (Ward, 1971; Cooper et al., 1972a). Radical complexation has, however, been suggested to explain the unusually weakly polarized signals observed in certain reactions of n-alkyl lithium reagents with n- or s-alkyl bromides (Ward, 1971). The reaction of n-butyl lithium with t-butyl bromide represents perhaps an intermediate case. With low concentrations of ether in the hydrocarbon solvent, the vinylic protons of the isobutene produced are polarized A/E indicating an encounter radical pair as precursor. Increasing the ether concentration reduces the intensity of the signal, eventually reversing its phase and bringing about an increase in intensity. The change in phase suggests that the olefin results from a geminate radical pair at the higher ether concentrations, but it occurs at just the point where it is known that the degree of aggregation of n-butyl lithium decreases from six to four. The hexamer, but not the tetramer, could perhaps give complexation.

Studies of alkyl halide–lithium alkyl reactions have been almost wholly concerned with proton polarization. The one exception to date is an investigation of ^{19}F polarization in the reaction of n-butyl lithium with p-fluorobenzyl chloride giving p,p'-difluorobibenzyl (A/E multiplet) and 1-fluoro-4-pentylbenzene (E/A) (Rakshys, 1970). Surprisingly ^{1}H-polarization is not observed.

2. *Reaction of organic halides with sodium naphthalene*

On the basis of reaction-product structures, it might be expected that the reactions of organic halides with sodium naphthalene (Scheme 9) might resemble mechanistically the reactions of organic halides with lithium alkyls. CIDNP studies have shown that they are in fact quite different, in particular in the mechanism by which polarization occurs. The observations are as follows (Garst et al., 1970).

   (i) Polarization of RR, RH and RX is maximal in magnetic fields of the order of 100 G, falling at lower and at higher fields.
   (ii) Polarized signals invariably occur as emissions.

(iii) Net polarizations are large at low fields, even though the difference in g-values between alkyl radicals and the naphthalene radical ion is small ($\Delta g = ca.\ 1\cdot 5 \times 10^{-4}$).

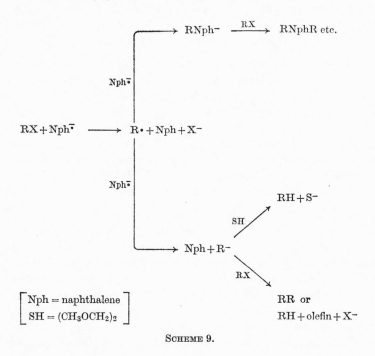

$$\begin{bmatrix} \text{Nph} = \text{naphthalene} \\ \text{SH} = (CH_3OCH_2)_2 \end{bmatrix}$$

SCHEME 9.

This type of polarization appears to be general in such systems and it has been proposed that it arises because of T_{-1}–S mixing. An interesting and important additional observation comes from a study of the reaction of isopropyl chloride with sodium naphthalene in a field of 60 G. Net emission in the products is predicted for T_{-1}–S mixing and multiplet polarization for T_0–S mixing. Since no multiplet polarization can be detected, it would seem that T_0–S mixing is suppressed in this system (Garst et al., 1971). Presumably, the naphthalene radical anion efficiently scavenges any radicals escaping the polarizing pair $\overline{R \cdot Nph^{\overline{\cdot}}}^T$ so that chemical partitioning, in which nuclear spin selection operates, cannot take place.

Somewhat similar observations have been made in the reaction of alkyl halides with sodium mirrors (the Wurtz reaction) in which alkyl coupling occurs. Thus, ethane formed on treatment of methyl iodide with sodium in a field of 20 G shows n.m.r. emission (Garst and Cox, 1970). The phase is consistent with polarization via T_{-1}–S mixing, provided that the initial radical pair is triplet (or an encounter pair). This is difficult to reconcile with suggestions that polarization occurs in the reaction $CH_3^- Na^+ + CH_3I \rightarrow CH_3.CH_3 + NaI$. Polarization is not observed in the reaction of 1,4-di-iodobutane with sodium to give cyclobutane, although polarization does appear when the naphthalene radical ion is used to bring about reaction. In the latter case the polarization is thought to occur by T_{-1}–S mixing in $\overline{I(CH_2)_3CH_2 \cdot {}^{-}Nph}^{T}$ pairs; in the former a 1,4-biradical is probably an intermediate and reasons have been given why polarization should not be possible in such species (Closs and Trifunac, 1969; but see also Closs, 1971a, b; Berson et al., 1971).

Net ^{19}F-polarization (A at 56·4 MHz) has been observed during the reaction of p-fluorobenzyl chloride with sodium naphthalene in tetrahydrofuran solution to give p, p'-difluorobibenzyl (Rakshys, 1971). The spin selection is believed to take place within a radical pair of the composition $\overline{pFC_6H_4CH_2 \cdot {}^{-}Nph}^{F}$. The final product then arises by a further electron transfer giving the benzyl anion, since magnesium bromide (which can trap the anion as a Grignard reagent) suppresses polarization. The polarization agrees in phase with predictions from Kaptein's rule, suggesting that T_0–S mixing is responsible in this reaction. On the other hand, at low fields, net emission occurs, and this may indicate T_{-1}–S mixing. Unusual field dependence of ^{19}F spectra has been reported in systems where there is no suggestion of T_1–S mixing (Bethell et al., 1972a) so that more detailed studies seem called for.

3. Organic derivatives of other metals

An almost perfect multiplet polarization (E/A) is observed for the α-protons of ethylmagnesium iodide when the reaction of ethyl iodide with magnesium is carried out in the probe of an n.m.r. spectrometer (Bodewitz et al., 1972; E. C. Ashby, personal communication). It has been argued that the polarization occurs during formation of the Grignard reagent and not as a result of the exchange reaction $RMgI + RI \rightleftarrows RI + RMgI$ [cf. exchange reactions involving lithium alkyls which do not engender polarization (see also, Ward et al., 1970)]. If this is so, is would lend support to inferences, drawn from the poor stereoselectivity of Grignard reactions involving substituted vinyl bromides, that radical intermediates are involved in the formation of vinylmagnesium halides (Méchin and Naulet, 1972). The multiplet nature of the observed polarization demands, of course, that $\Delta g \approx 0$ in the radical pair leading to the alkylmagnesium halide: this seems surprising for, say, $\overline{CH_3CH_2 \cdot \cdot MgBr}$.

Polarization also occurs in coupling and disproportionation reactions of Grignard reagents with alkyl halides. The vinyl protons of isobutene produced in the reaction of t-butylmagnesium chloride with t-butyl bromide show A/E polarization as do the methyl protons of isobutane (Ward *et al.*, 1970). Similar results arise in the reaction of diethyl-magnesium with organic halides (Kasukhin *et al.*, 1972).

CIDNP has also been reported in reactions of organomercurials. Emission is observed from the coupling product of *p*-methylbenzyl-mercuric bromide and triphenylmethyl bromide (Beletskaya *et al.*, 1971), while thermolysis of organomercury derivatives of tin such as t-$C_4H_9HgSn(CH_3)_3$ gave mixtures of isobutene and isobutane (by disproportionation of uncorrelated pairs of t-butyl radicals) showing A/E polarization (Mitchell, 1972).

F. *Ylid and Related Molecular Rearrangements*

A fundamental problem associated with mechanisms of rearrangement reactions is concerned with whether the transformation occurs in a concerted or stepwise manner. The alternatives are illustrated in a general way in equation (62).

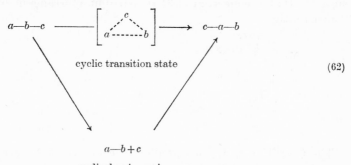

$$a\text{---}b\text{---}c \longrightarrow \left[\begin{array}{c} c \\ a\text{------}b \end{array} \right] \longrightarrow c\text{---}a\text{---}b \tag{62}$$

cyclic transition state

$a\text{---}b + c$

radical or ion pair

CIDNP represents a particularly direct and sensitive method for detecting stepwise rearrangements involving radical pairs. The method has had some notable successes, but it is worth stressing that the results of CIDNP studies need to be interpreted particularly cautiously in rearrangement reactions. Radical mechanisms can represent a very minor side reaction competing with a much more important non-radical route (Iwamura *et al.*, 1970; Jacobus, 1970). On the other hand, failure to observe polarization need not rule out the radical pathway for re-arrangement. In general, the precursor of the radical pair is a singlet molecule, so that the radical pair also is formed initially in the singlet state. Unless the lifetime of the radical pair is long enough for T_0–S mixing to occur, rearrangement could occur without polarization. This is particularly likely in those cases when the components of the radical pair are part of the same intermediate (i.e. a biradical), although $T_{\pm 1}$–S mixing could lead to polarization in some cases.

CIDNP has proved useful in testing the predictions of the theory of conservation of orbital symmetry in rearrangement reactions. So far, interest has centred largely on 4-electron systems where the concerted (sigmatropic) reactions are designated as being of order [1,3] for electrically neutral reactants, [1,4] for cationic and [1,2] for anionic reactants. For migration of hydrogen, the thermally allowed concerted process involves an antarafacial migration; for migrating groups capable of utilizing orbitals of p-symmetry, suprafacial migration with inversion of configuration is also allowed, antarafacial migration occurring with retention (Woodward and Hoffmann, 1970; Gilchrist and Storr, 1972).

A radical pair pathway has been identified for the conversion of **29** into **30** as indicated by the E/A multiplet due to the A_2B_2 system in the amine side chain at δ 2·5–3·0 ppm (Baldwin and Brown, 1969; for CIDNP in the formation of a methylenecyclohexadiene, see Bethell *et al.*, 1972c). The low enhancement factor (5–20) at maximum polarization suggests that a concerted [1,3] sigmatropic rearrangement may occur concurrently.

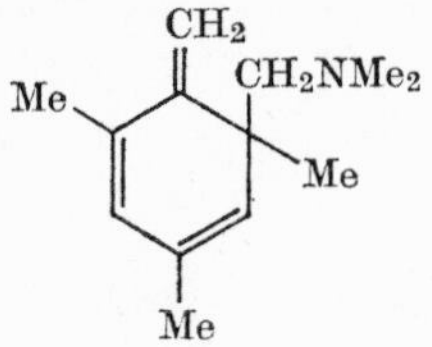

(29) 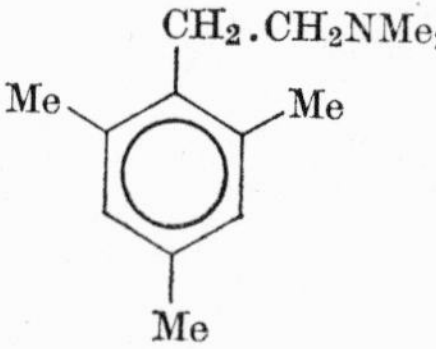   (30)

The transition state for the allowed concerted 1,2-rearrangement of anionic systems is difficult to achieve. The 1,2-shift of an alkyl group in an all-carbon system has been reported but the stereochemistry is not known (Grovenstein and Wentworth, 1967). Much more common are 1,2-shifts in heteronuclear systems, the so-called ylid rearrangements, such as the Stevens rearrangement (**31** → **32**). These often occur with retention of configuration in the migrating group, which is inconsistent

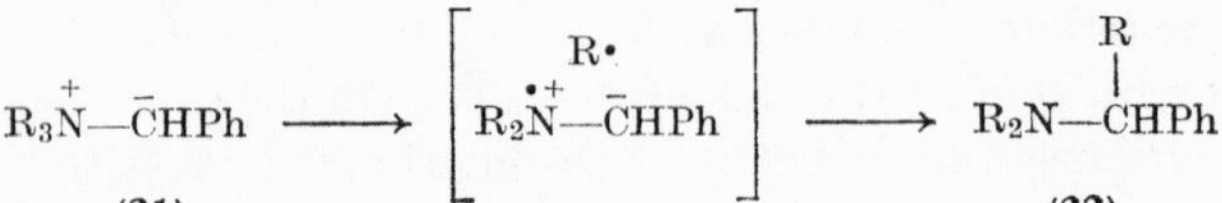

(31)   (32)

with the predictions from orbital symmetry considerations. On the other hand, retention in the migrating group seems inconsistent with its separation as a free entity. Many examples of CIDNP in ^{1}H n.m.r. spectra of reacting systems have been described (Schöllkopf, 1970, 1971). These reports are summarized in Table 1. Clearly, radical pair pathways are available for rearrangement and, in the case of **33**, 36% retention of configuration in the migrating α-deuteriobenzyl group in product **34** has been reported (Baldwin *et al.*, 1970b). What is not proven, however, is that the polarized molecules have the retained configuration. Further investigation is needed.

The occurrence of characteristic by-products can be taken as indicative of a radical pair mechanism in rearrangements. Thus the observation

$$\text{Ph}\overset{-}{\text{C}}\text{O}\overset{+}{\text{C}}\text{HS(Me)CHDPh} \qquad\qquad \text{PhCOCH(CHDPh)SMe}$$

$$(33) \qquad\qquad\qquad\qquad (34)$$

of bibenzyl in the rearrangement of the sulphonium ylid **35** is suggestive of separation of the migrating benzyl group as a free radical followed by dimerization (Schöllkopf *et al.*, 1970). However, the benzylic protons show enhanced absorption in the n.m.r. spectrum of the reaction mixture. This is inconsistent with formation of bibenzyl by reaction of a uncorrelated pair of benzyl radicals: a memory effect seems indicated.

When the migrating group is allyl, an additional concerted ([2,3] sigmatropic) pathway for rearrangement becomes available. In this an allylic shift must also occur. Nevertheless, the radical pathway is not always excluded. For example, rearrangement of ylids such as **36** (R = CH$_3$.CO) leads to product **37** (R = CH$_3$.CO) in which the allylic protons adjacent to the amido-nitrogen atom appear in emission (D. G. Morris, 1969). No polarization is observed in the much readier

$$\begin{array}{ccc}
\text{CH}_2\text{Ph} & \text{R}\overset{-}{\text{N}}-\overset{+}{\text{N}}(\text{CH}_3)_2 & \text{R}-\text{N}-\text{N}(\text{CH}_3)_2 \\
| & \diagdown & \diagup \\
\text{CH}_3-\text{S}-\overset{-}{\text{C}}\text{HCOPh} & \text{CH}_2 & \text{CH}_2 \\
\overset{+}{} & \diagup & \diagdown \\
& (\text{CH}_3)_2\text{C}=\text{CH} & \text{CH}=\text{C}(\text{CH}_3)_2 \\
(35) & (36) & (37)
\end{array}$$

$$\begin{array}{c}
\text{RN}-\text{N}(\text{CH}_3)_2 \\
| \\
(\text{CH}_3)_2\text{C}\diagdown \\
\phantom{(\text{CH}_3)_2\text{C}}\text{CH}=\text{CH}_2 \\
(38)
\end{array}$$

rearrangement of **36** (R = H) giving the allylic rearrangement product **38**, and it has been suggested that the full negative charge localized on nitrogen is necessary to ensure a concerted mechanism (Baldwin *et al.*, 1970a).

TABLE 1
CIDNP in Ylid and Related Rearrangements

Type	Reactant	Product	Polarization	Reference
N—C	PhCH₂N⁺Me₂C⁻HCOPh	Me₂N.CH(COPh)CH₂Ph	E	Schöllkopf et al., 1969b
	PhN⁺Me(CH₂Ph)C⁻HPh	PhNMe.CHPh.CH₂Ph	A/E; E/A	Lepley, 1969b
	Ph.N⁺Me₂C⁻HPh	PhNMe.CHPh.CH₃	A/E	Lepley et al., 1971
N—N	p-O₂N.C₆H₄.CH₂N⁺Me₂N⁻.COCH₃	Me₂N.N(COCH₃)CH₂.C₆H₄NO₂-p	E	Jemison and Morris, 1969

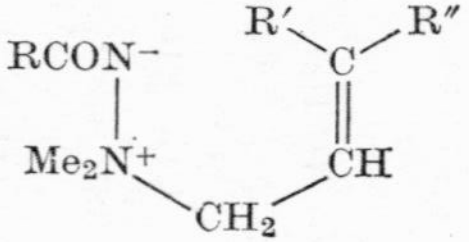

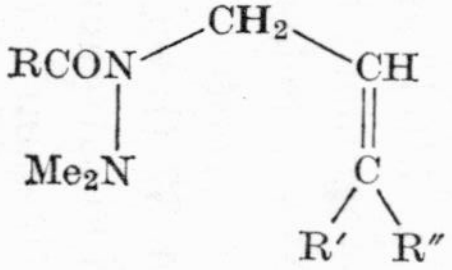

$$RCON^-\!\!-\!N^+Me_2(CH_2)(R'C=CHR'') \longrightarrow RCON(CH_2)(Me_2N)(CH=CR'R'')$$

R	R′	R″	Polarization	Reference
CH₃	H	CH₃	E	Baldwin et al., 1970a
CH₃	CH₃	CH₃		D. G. Morris, 1969; Baldwin et al., 1970a
CH₃	H	C₆H₅		Baldwin et al., 1970a
C₆H₅	H	H		Baldwin et al., 1970a

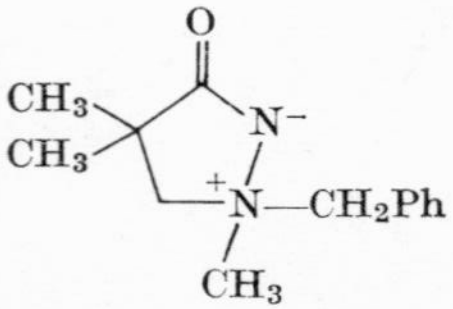

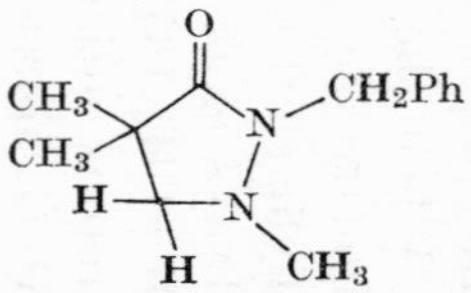

(ring structure: 2,2-dimethyl-3-oxo-1-CH₂Ph-1-N⁺-2-N⁻-CH₃ pyrazolidinone) → (rearranged ring with CH₂Ph and CH₃), E, Benecke and Wikel, 1971; D. G. Morris, 1971b

5

	Reactant	Product		Reference
N—O	$PhCH_2\overset{+}{N}(O^-)(Ph)CH_3$	$PhN(CH_3)OCH_2Ph$	E; E	Ostermann and Schöllkopf, 1970
	$PhCH_2\overset{+}{N}(O^-)(CH_3)_2$	$(CH_3)_2NOCH_2Ph$	E; E	Lepley et al., 1970
	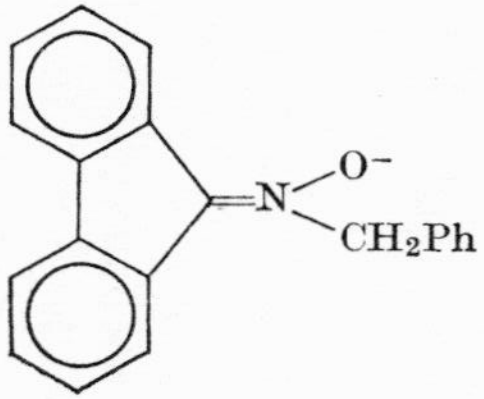 fluorenylidene-$N(O^-)$—CH_2Ph	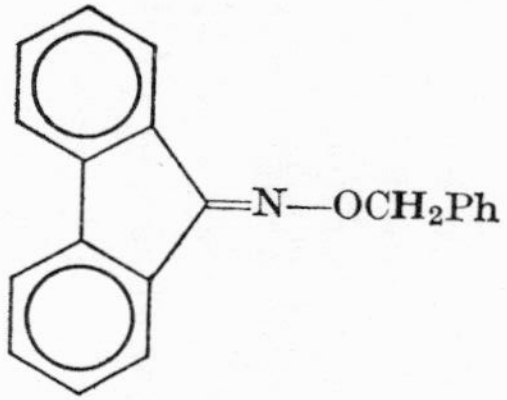 fluorenylidene=N—OCH_2Ph	E	Morris, 1971a
O—C	$PhCHOC(CH_3)_3Li^+$	$(CH_3)_3C \cdot CHPh(O^-)Li^+$	E	Lansbury, 1970
S—C	$Ph\overset{+}{CH_2}S(CH_3)\overset{-}{CH}COPh$	$\begin{cases} CH_3SCH(COPh)CH_2Ph \\ PhCH_2CH_2Ph \end{cases}$	A; E A	Schöllkopf et al., 1969a 1970
	$PhCHD\overset{+}{S}(CH_3)\overset{-}{CH}COPh$	$CH_3SCH(COPh)CHDPh$	A, E	Baldwin et al., 1970b
	$Ph\overset{+}{CH_2}S(Ph)\overset{-}{CH}Ph$	$PhSCH(Ph)CH_2Ph$	A; E	Iwamura et al., 1971a
O—S	$CH_3O \cdot CO \cdot CH_2\overset{+}{S}(C_2H_5)\overset{-}{C}(CO \cdot OCH_3)_2$	$C_2H_5SC(CO \cdot OCH_3)_2CH_2CO \cdot OCH_3$	E	Ando et al., 1972
	$p\text{-}CH_3 \cdot C_6H_4 \cdot CO \cdot N(CH_3)O \cdot CS \cdot NMe_2$	$p\text{-}CH_3 \cdot C_6H_4 \cdot CO \cdot N(CH_3)S \cdot CONMe_2 +$	A	Ankers et al., 1972
		$p\text{-}CH_3 \cdot C_6H_4 \cdot CONHCH_3$	E	

CIDNP has been sought in other molecular rearrangements [e.g., the benzidine rearrangement (Banthorpe and Winter, 1972)] but so far without success. The conversion of **39** into **40** is noteworthy since it is related to [3,3]sigmatropic rearrangements, such as the Cope rearrangement, but involves allylic rearrangement in only one half of the molecule. The absence of observable polarization has been taken to indicate an ion-pair rather than a radical-pair mechanism for this stepwise rearrangement (Wigfield *et al.*, 1972). Polarization observed in the

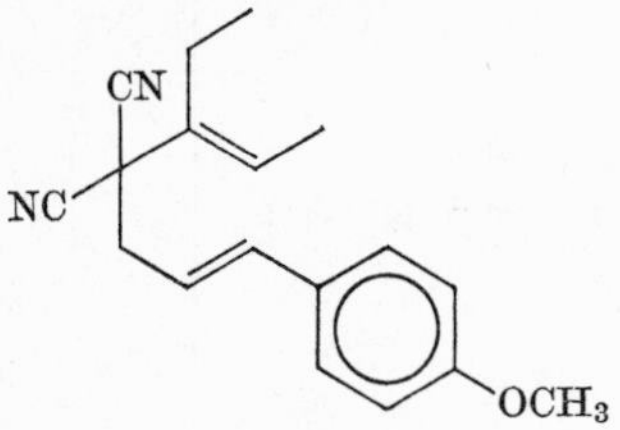

(39) 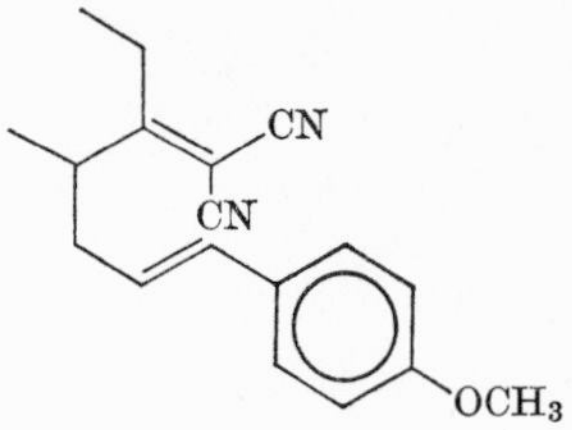   (40)

reaction of α-picoline- and related N-oxides (e.g. **41**) with acetic anhydride, giving acetoxymethyl derivatives of the heterocycle (e.g. **42**), appears to be associated exclusively with by-products; the rearrangement product shows no polarization, suggesting a concerted mechanism for the main reaction (Iwamura *et al.*, 1970a, c; see also, Kawazoe and Araki, 1971; Atlanti *et al.*, 1971; for a review of earlier work suggesting a radical pair mechanism, see Shine, 1967).

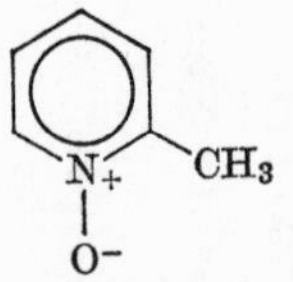

(41) 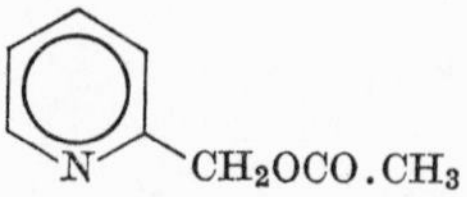   (42)

VI. CHEMICALLY INDUCED DYNAMIC ELECTRON SPIN POLARIZATION (CIDEP)

Radicals escaping from a radical pair become uncorrelated as J_{ee} approaches zero. In the free (doublet) state they are detectable by e.s.r. spectroscopy. However, just as polarization of nuclear spins can occur in the radical pair, so polarization of electron spins can be produced. Provided that electron spin-lattice relaxation and free radical scavenging processes do not make the lifetime of the polarized radicals too short, polarized e.s.r. signals should be detectable. Indeed, such CIDEP was first observed for hydrogen atoms formed in the radiolysis of hydrocarbons long before the phenomenon of CIDNP was recognized (Fessenden and Schuler, 1963). Since then several further examples have been reported in radiolytic and photolytic experiments (Livingston and Zeldes, 1966, 1970: Smaller *et al.*, 1968, 1971; Krusic and Kochi, 1968; Fischer and Paul, 1970; Zeldes and Livingston, 1970; Atkins *et al.*, 1970a; Glarum and Marshall 1970; Veeman and van der Waals, 1970; Eiben and Fessenden, 1971; Neta *et al.*, 1971). Only with the development of the theory of CIDNP was an acceptable theory of CIDEP produced (Fischer and Bargon, 1969; Kaptein and Oosterhoff, 1969a; see also Fischer, 1970b; Glarum and Marshall, 1970; Atkins *et al.*, 1971; Hutchinson *et al.*, 1972). Again the most successful theoretical approach has been through the random-walk or diffusion model of radical-pair behaviour (Adrian, 1971b, 1972).

CIDEP usually results from T_0–S mixing in radical pairs, although T_{-1}–S mixing has also been considered (Atkins *et al.*, 1971, 1973). The time development of electron-spin state populations is a function of the electron Zeeman interaction, the electron-nuclear hyperfine interaction, the electron–electron exchange interaction, together with spin-rotational and orientation dependent terms (Pedersen and Freed, 1972). Electron spin lattice relaxation ($T_{1e} = 10^{-5}$ to 10^{-6} sec) is normally slower than the polarizing process.

Pulse techniques, coupled with the observation of the decay of enhancement (Atkins *et al.*, 1970a, b; Glarum and Marshall, 1970; Smaller *et al.*, 1971) constitute the most sensitive procedure for detecting CIDEP. Both net and multiplet polarization have been described. As with CIDNP, the former is believed to arise essentially from the Zeeman interaction and the latter from the hyperfine term. Qualitative rules analogous to Kaptein's rules should be capable of development.

As yet there has been little exploitation of CIDEP. The phenomenon is potentially very valuable in the study of transient radicals and should complement CIDNP.

VII. Prospects

The strenuous efforts made in the study of CIDNP over the past few years have produced a basic quantitative theory of the effect. This has been extensively tested, chiefly in well-understood areas of radical chemistry, and found to be satisfactory. There still remains room for theoretical development and refinement, however, but the time is now ripe for the exploitation of the phenomenon in chemical studies. The occurrence of chemical as well as spectroscopic consequences of T–S mixing is now beginning to be considered (see, for example, Lawler and Evans, 1971).

In the area of spectroscopy, the application, with computer assistance, of the theory to observed spectra should provide much readier access to magnetic resonance parameters (e.g., radical g-factors, absolute signs of a- and J-values) than has been available hitherto. Moreover, for radical parameters, CIDNP is much more versatile than e.s.r. spectroscopy, since it can readily handle even very reactive radicals.

The potential of CIDNP in mechanistic studies of organic reactions is only just beginning to be developed. Of necessity, the early studies have been to a large extent tests of the radical-pair theory where new mechanistic information was scarcely to be expected. Nevertheless, as shown in Section V, CIDNP has produced stimulating new insights into a wide variety of reactions, challenging in some areas earlier views based on conventional mechanistic investigations. This process can be expected to continue. The technique complements the spin-trapping procedure used in e.s.r. spectroscopy, since the effect is directly related to the chemical reactions undergone by the radicals as well as the structure of the radicals.

More systematic study of the dynamics of radicals in solution should now be possible using CIDNP. Investigations so far reported have indicated that the rates of very rapid chemical reactions and other dynamic processes undergone by radicals can be measured in a crude way; greater refinement should be possible. Special effects have been predicted for reactions in thin films (Deutch, 1972). Moreover the time-scale of polarization is such that the technique may prove capable of throwing new light on the dynamics of excited states.

Radical chemistry has undergone something of a renaissance in recent years. The phenomenon of CIDNP has played an important part in this. The growing interest in the role of radical processes in biological systems may stimulate the application of CIDNP in even wider fields in the future. The development of a practical device for radiofrequency amplification by the stimulated emission of radiation (RASER) may well be one such application.

REFERENCES

Abram, I. I., Milne, G. S., Solomon, B. S., and Steel, C. (1969). *J. Am. Chem. Soc.* **91**, 1220.

Adrian, F. J. (1970). *J. Chem. Phys.* **53**, 3374.

Adrian, F. J. (1971a). *J. Chem. Phys.* **54**, 3912.

Adrian, F. J. (1971b). *J. Chem. Phys.* **54**, 3918.

Adrian, F. J. (1972). *J. Chem. Phys.* **57**, 5107.

Ando, W., Yagihara, T., Tozune, S., Imai, I., Suzuki, J., Toyama, T., Nakaido, S., and Migita, T. (1972). *J. Org. Chem.* **37**, 1721.

Ankers, W. B., Brown, C., Hudson, R. F., and Lawson, A. J. (1972). *Chem. Comm.* 935.

Atkins, P. W., Buchanan, I. C., Gurd, R. C., McLauchlan, K. A., and Simpson, A. F. (1970a). *Chem. Comm.* 513.

Atkins, P. W., Gurd, R. C., McLauchlan, K. A., and Simpson, A. F. (1971). *Chem. Phys. Lett.* **8**, 55.

Atkins, P. W., McLauchlan, K. A., and Percival, P. W. (1973). *Chem. Comm.* 121.

Atkins, P. W., McLauchlan, K. A., and Simpson, A. F. (1970b). *J. Phys. (E)* **3**, 547.

Atlanti, P., Biellmann, J., Brière, R., Lemaire, H., and Rassat, A. (1971). *Ind. Chim. Belge* **36**, 1066.

Baldwin, J. E., and Brown, J. E. (1969). *J. Am. Chem. Soc.* **91**, 3647.

Baldwin, J. E., Brown, J. E., and Cordell, R. W. (1970a). *Chem. Comm.* 31.

Baldwin, J. E., Erickson, W. F., Hackler, R. E., and Scott, R. M. (1970b). *Chem. Comm.* 576.

Bamford, C. H., and Casson, J. E. (1969). *Proc. Roy. Soc.* **A312**, 163.

Banthorpe, D. V., and Winter, J. G. (1972). *J.C.S. Perkin II* 868.

Bargon, J. (1971a). *J. Polym. Sci. Part B* **9**, 681.

Bargon, J. (1971b). *J. Am. Chem. Soc.* **93**, 4630.

Bargon, J. (1971c). *Ind. Chim. Belge* **36**, 1061.

Bargon, J., and Fischer, H. (1967). *Z. Naturforsch.* **22a**, 1556.

Bargon, J., and Fischer, H. (1968a). *Z. Naturforsch.* **23a**, 2109.

Bargon, J., and Fischer, H. (1968b). *Z. Naturforsch.* **23a**, 2125.

Bargon, J., Fischer, H., and Johnsen, V. (1967). *Z. Naturforsch.* **22a**, 1551.

Bartlett, P. D., and Engel, P. S. (1968). *J. Am. Chem. Soc.* **90**, 2960.

Beletskaya, I. P., Vol'eva, V. B., Rykov, S. V., Buchachenko, A. L., and Kessenikh, A. V. (1971). *Izv. Akad. Nauk SSSR, Ser. Khim.* 454.

Benecke, H. P., and Wikel, J. H. (1971). *Tetrahedron Lett.* 3479.

Berger, S., Hauff, S., Niederer, P., and Rieker, A. (1972). *Tetrahedron Lett.* 2581.

Berson, J. A., Bushby, R. J., McBride, J. M., and Tremelling, M. (1971). *J. Am. Chem. Soc.* **93**, 1544.

Bethell, D. (1969). *Adv. Phys. Org. Chem.* **7**, 153.

Bethell, D., Whittaker, D., and Callister, J. D. (1965). *J. Chem. Soc.* 2466.

Bethell, D., Stevens, G., and Tickle, P. (1970). *Chem. Comm.* 792.

Bethell, D., Brinkman, M. R., and Hayes, J. (1972a). *Chem. Comm.* 475.

Bethell, D., Brinkman, M. R., and Hayes, J. (1972b). *Chem. Comm.* 1323.

Bethell, D., Brinkman, M. R., and Hayes, J. (1972c). *Chem. Comm.* 1324.

Bilevich, K. A., Bubnov, N. N., Medvedev, B. Y., Okhlobystin, O. Y., and Ermanson, L. V. (1970). *Dokl. Akad. Nauk SSSR.* **193**, 583.

Bilevich, K. A., Bubnov, N. N., Ioffe, N. T., Kalinkin, M. I., Okhlobystin, O. Y., and Petrovskii, P. V. (1971). *Izv. Akad. Nauk SSSR, Ser. Khim.* 1707.

Blank, B., and Fischer, H. (1971a). *Ind. Chim. Belge* **36**, 1075.

Blank, B., and Fischer, H. (1971b). *Helv. Chim. Acta* **54**, 905.

Blank, B., Mennitt, P. G., and Fischer, H. (1971). "Special Lectures presented at the XXIIIrd International Congress of Pure and Applied Chemistry, Boston," Vol. 4, Butterworths, London, p. 1.

Bodewitz, H. W. H. J., Blomberg, C., and Bickelhaupt, F. (1972). *Tetrahedron Lett.* 281.

Brinkman, M. R., Bethell, D., and Hayes, J. (1973a). *Tetrahedron Lett.* 989.

Brinkman, M. R., Bethell, D., and Hayes, J. (1973b). *J. Chem. Phys.* in press.

Bryce-Smith, D. (1956). *J. Chem. Soc.* 1603.

Bubnov, N. N., Bilevitch, K. A., Poljakova, L. A., and Okhlobystin, O. Y. (1972). *Chem. Comm.* 1058.

Buchachenko, A. L., and Zhidomirov, G. M. (1971). *Usp. Khim.* **40**, 1729; *Russ. Chem. Rev.* **40**, 801.

Buchachenko, A. L., Rykov, S. V., Kessenikh, A. V., and Bylina, G. S. (1970a). *Dokl. Akad. Nauk SSSR* **190**, 839.

Buchachenko, A. L., Rykov, S. V., and Kessenikh, A. V. (1970b). *Zhur. Fiz. Khim.* **44**, 876; *Russ. J. Phys. Chem.* **44**, 488.

Buchachenko, A. L., Kessenikh, A. V., and Rykov, S. V. (1970c). *Zh. Eksp. Teor. Fiz.* **58**, 766.

Cadogan, J. I. G. (1970). *In* "Essays on Free Radical Chemistry", Chemical Society Special Publication, No. 24, p. 71.

Carrington, A., and McLachlan, A. D. (1967). "Introduction to Magnetic Resonance", Harper and Row, New York.

Charlton, J. L., and Bargon, J. (1971). *Chem. Phys. Lett.* **8**, 442.

Clark, W. G., Setser, D. W., and Siefert, E. E. (1970). *J. Phys. Chem.* **74**, 1670.

Closs, G. L. (1968). *Topics in Stereochem.* **3**, 193.

Closs, G. L. (1969). *J. Am. Chem. Soc.* **91**, 4552.

Closs, G. L. (1971a). "Special Lectures presented at the XXIIIrd International Congress of Pure and Applied Chemistry, Boston", Vol. 4, Butterworths, London, p. 19.

Closs, G. L. (1971b). *J. Am. Chem. Soc.* **93**, 1546.

Closs, G. L. (1971c). *Ind. Chim. Belge* **36**, 1064.

Closs, G. L., and Closs, L. E. (1969a). *J. Am. Chem. Soc.* **91**, 4549.

Closs, G. L., and Closs, L. E. (1969b). *J. Am. Chem. Soc.* **91**, 4550.

Gloss, G. L., and Doubleday, C. E. (1972). *J. Am. Chem. Soc.* **94**, 9248.

Closs, G. L., and Paulson, D. R. (1970). *J. Am. Chem. Soc.* **92**, 7229.

Closs, G. L., and Trifunac, A. D. (1969). *J. Am. Chem. Soc.* **91**, 4554.

Closs, G. L., and Trifunac, A. D. (1970a). *J. Am. Chem. Soc.* **92**, 2183.

Closs, G. L., and Trifunac, A. D. (1970b). *J. Am. Chem. Soc.* **92**, 2186.

Closs, G. L., and Trifunac, A. D. (1970c). *J. Am. Chem. Soc.* **92**, 7227.

Closs, G. L., Doubleday, C. E., and Paulson, D. R. (1970). *J. Am. Chem. Soc.* **92**, 2185.

Cocivera, M. (1968). *J. Am. Chem. Soc.* **90**, 3261.

Cocivera, M., and Roth, H. D. (1970). *J. Am. Chem. Soc.* **92**, 2573.

Cocivera, M., and Trozzolo, A. M. (1970). *J. Am. Chem. Soc.* **92**, 1772.

Cooper, R. A., Lawler, R. G., and Ward, H. R. (1972a). *J. Am. Chem. Soc.* **94**, 545.

Cooper, R. A., Lawler, R. G., and Ward, H. R. (1972b). *J. Am. Chem. Soc.* **94**, 552.

Cubbon, R. C. P. (1970). *Progr. React. Kinetics* **5**, 29.

den Hollander, J. A. (1971). *Ind. Chim. Belge* **36**, 1083.

den Hollander, J. A., Kaptein, R., and Brand, P. A. T. M. (1971). *Chem. Phys. Lett.* **10**, 430.

DeTar, D. F. (1967). *J. Am. Chem. Soc.* **89**, 4058.

Deutch, J. M. (1972). *J. Chem. Phys.* **56**, 6076.

Dombchik, S. A. (1969). Ph.D. Thesis, University of Illinois, quoted by Cooper *et al.* (1972a).

DoMinh, T. (1971). *Ind. Chim. Belge* **36**, 1080.

Eiben, E., and Fessenden, R. W. (1971). *J. Phys. Chem.* **75**, 1186.

Fahrenholtz, S. R., and Trozzolo, A. M. (1971). *J. Am. Chem. Soc.* **93**, 251.

Fahrenholtz, S. R., and Trozzolo, A. M. (1972). *J. Am. Chem. Soc.* **94**, 282.

Ferruti, P., Gill, D., Klein, M. P., Wang, H. H., Entine, G., and Calvin, M. (1970). *J. Am. Chem. Soc.* **92**, 3704.

Fessenden, R. W., and Schuler, R. H. (1963). *J. Chem. Phys.* **39**, 2147.

Fischer, H. (1969). *J. Phys. Chem.* **73**, 3834.

Fischer, H. (1970a). *Z. Naturforsch.* **25a**, 1957.

Fischer, H. (1970b). *Chem. Phys. Lett.* **4**, 611.

Fischer, H. (1971a). *Fortschr. Chem. Forsch.* **24**, 1; *Topics in Current Chemistry*, **24**, 1.
Fischer, H. (1971b). *Ind. Chim. Belge* **36**, 1054.
Fischer, H., and Bargon, J. (1969). *Accounts Chem. Res.* **2**, 110.
Fischer, H., and Lehnig, M. (1971). *J. Phys. Chem.* **75**, 3410.
Fischer, H., and Paul, H. (1970). *Z. Naturforsch.* **25a**, 443.
Garst, J. F., and Cox, R. H. (1970). *J. Am. Chem. Soc.* **92**, 6389.
Garst, J. F., Cox, R. H., Barbas, J. T., Roberts, R. D., Morris, J. I., and Morrison, R. C. (1970). *J. Am. Chem. Soc.* **92**, 5761.
Garst, J. F., Barton, F. E., and Morris, J. I. (1971). *J. Am. Chem. Soc.* **93**, 4310.
Gilchrist, T. L., and Storr, R. C. (1972). "Organic Reactions and Orbital Symmetry", Cambridge University Press, London
Glarum, S. H., and Marshall, J. H. (1970). *J. Chem. Phys.* **52**, 5555.
Grovenstein, E., and Wentworth, G. (1967). *J. Am. Chem. Soc.* **89**, 8152, 2348.
Hargis, J. H., and Shevlin, P. B. (1973). *Chem. Comm.* 179.
Hausser, K. H., and Stehlik, D. (1968). *Adv. Mag. Res.* **3**, 79.
Heesing, A., and Kaiser, B. H. (1970). *Tetrahedron Lett.* 2845.
Herring, C., and Flicker, M. (1964). *Phys. Rev.* **134**, A362.
Hiatt, R. (1971). *In* "Organic Peroxides" (D. S. Swern, ed.), Vol. 2, Wiley-Interscience, New York, Ch. 8.
Hirota, H., and Weissman, S. I. (1964). *J. Am. Chem. Soc.* **86**, 2538.
Hoffmann, R. W., and Guhn, G. (1967). *Chem. Ber.* **100**, 1474.
Hollaender, J., and Neumann, W. P. (1970). *Angew. Chem.* **82**, 813; *Int. Ed.* **9**, 804.
Hollaender, J., and Neumann, W. P. (1971). *Ind. Chim. Belge* **36**, 1072.
Hutchinson, D. A., Wong, S. K., Colpa, J. P., and Wan, J. K. S. (1972). *J. Chem. Phys.* **57**, 3308.
Ignatenko, A. V., Kessenikh, A. V., and Rykov, S. V. (1971). *Izv. Akad. Nauk SSSR, Ser. Khim.* 1368.
Itoh, K., Hoyoshi, H., and Nagakura, S. (1969). *Mol. Phys.* **17**, 561.
Iwamura, H. (1971). *Yuki Gosei Kagaku Kyokai Shi* **29**, 15.
Iwamura, H., and Iwamura, M. (1970). *Tetrahedron Lett.* 3723.
Iwamura, H., Iwamura, M., Nishida, T., and Miura, I. (1970a). *Bull. Chem. Soc. Jap.* **43**, 1914.
Iwamura, H., Iwamura, M., Tamura, M., and Shiomi, K. (1970b). *Bull. Chem. Soc. Jap.* **43**, 3638.
Iwamura, H., Iwamura, M., Nishida, I., and Miura, I. (1970c). *Tetrahedron Lett.* 3117.
Iwamura, H., Iwamura, M., Nishida, T., Yoshida, M., and Nakayama, J. (1971a). *Tetrahedron Lett.* 63.
Iwamura, H., Iwamura, M., Sato, S., and Kushida, K. (1971b). *Bull. Chem. Soc. Jap.* **44**, 876.
Jacobus, J. (1970). *Chem. Comm.* 709.
Jemison, R. W., and Morris, D. G. (1969). *Chem. Comm.* 1226.
Kaplan, M. L., and Roth, H. D. (1972). *Chem. Comm.* 970.
Kaptein, R. (1968). *Chem. Phys. Lett.* **2**, 261.
Kaptein, R. (1971a). *Chem. Comm.* 732.
Kaptein, R. (1971b). "Chemically Induced Dynamic Nuclear Polarization", Thesis, Leiden.
Kaptein, R. (1972a). *J. Am. Chem. Soc.* **94**, 6251.
Kaptein, R. (1972b). *J. Am. Chem. Soc.* **94**, 6262.
Kaptein, R., and den Hollander, J. A. (1972). *J. Am. Chem. Soc.* **94**, 6269.
Kaptein, R., and Oosterhoff, L. J. (1969a). *Chem. Phys. Lett.* **4**, 195.
Kaptein, R., and Oosterhoff, L. J. (1969b). *Chem. Phys. Lett.* **4**, 214.

126   D. BETHELL AND M. R. BRINKMAN

Kaptein, R., Brokken-Zijp, J., and de Kanter, F. J. J. (1972). *J. Am. Chem. Soc.* **94**, 6280.
Kaptein, R., den Hollander, J. A., Antheunis, D., and Oosterhoff, L. J. (1970). *Chem. Comm.* 1687.
Kaptein, R., Frater-Schröder, M., and Oosterhoff, L. J. (1971). *Chem. Phys. Lett.* **12**, 16.
Kaptein, R., Verheus, F. W., and Oosterhoff, L. J. (1971). *Chem. Comm.* 877.
Kasukhin, L. F., Ponomarchuk, M. P., and Kalinin, V. N. (1970). *Zh. Org. Khim.* **6**, 2531.
Kasukhin, L. F., Ponomarchuk, M. P., and Buteiko, Z. F. (1972). *Zh. Org. Khim.* **8**, 665.
Kawazoe, Y., and Araki, M. (1971). *Chem. Pharm. Bull.* **19**, 1278.
Kessenikh, A. V., Rykov, S. V., and Buchachenko, A. L. (1970). *Zh. Eksp. Teor. Fiz.* **59**, 387.
Kessenikh, A. V., Rykov, S. V., and Yankelevich, A. Z. (1971). *Chem. Phys. Lett.* **9**, 347.
Kobrina, L. S., Vlasova, L. V., and Mamatyuk, V. I. (1972). *Sib. Otd. Akad. Nauk SSSR, Ser. Khim. Nauk Izv.* 92.
Koenig, T., and Mabey, W. R. (1970). *J. Am. Chem. Soc.* **92**, 3804.
Krusic, P. J., and Kochi, J. K. (1968). *J. Am. Chem. Soc.* **90**, 7155.
Lane, A. G., Rüchardt, C., and Werner, R. (1969). *Tetrahedron Lett.* 3213.
Lansbury, P. T. (1970). Unpublished experiments, quoted by Schöllkopf (1970).
Lawler, R. G. (1967). *J. Am. Chem. Soc.* **89**, 5519.
Lawler, R. G. (1972). *Accounts Chem. Res.* **5**, 25.
Lawler, R. G., and Evans, G. T. (1971). *Ind. Chim. Belge* **36**, 1087.
Lawler, R. G., Ward, H. R., Allen, R. B., and Ellenbogen, P. E. (1971). *J. Am. Chem. Soc.* **93**, 789.
Lehnig, M., and Fischer, H. (1969). *Z. Naturforsch.* **24a**, 1771.
Lehnig, M., and Fischer, H. (1970). *Z. Naturforsch.* **25a**, 1963.
Lepley, A. R. (1968). *J. Am. Chem. Soc.* **90**, 2710.
Lepley, A. R. (1969a). *J. Am. Chem. Soc.* **92**, 749.
Lepley, A. R. (1969b). *Chem. Comm.* 1460.
Lepley, A. R. (1969c). *Chem. Comm.* 64.
Lepley, A. R., and Landau, R. L. (1969). *J. Am. Chem. Soc.* **92**, 748.
Lepley, A. R., Cook, P. M., and Willard, G. F. (1970). *J. Am. Chem. Soc.* **92**, 1101.
Lepley, A. R., Becker, R. H., and Giumanini, A. G. (1971). *J. Org. Chem.* **36**, 1222.
Levin, Y. A., Il'yasov, A. V., Pobedimskii, D. G., Gol'dfarb, E. I., Saidashev, I. I., and Samitov, Y. Y. (1970). *Izv. Akad. Nauk SSSR, Ser. Khim.* 1680.
Levit, A. F., Gragerov, I. P., and Buchachenko, A. L. (1971). *Dokl. Akad. Nauk SSSR* **201**, 897.
Levit, A. F., Buchachenko, A. L., Kiprianova, L. A., and Gragerov, I. P. (1972). *Dokl. Akad. Nauk SSSR* **203**, 628.
Lippmaa, E. T., Pekhk, T. I., Buchachenko, A. L., and Rykov, S. V. (1970a). *Chem. Phys. Lett.* **5**, 521.
Lippmaa, E. T., Pekhk, T. I., Buchachenko, A. L., and Rykov, S. V. (1970b). *Dokl. Akad. Nauk SSSR* **195**, 632.
Lippmaa, E. T., Pekhk, T. I., and Saluvere, T. (1971). *Ind. Chim. Belge* **36**, 1070.
Livingston, R., and Zeldes, H. (1966). *J. Am. Chem. Soc.* **88**, 4333.
Livingston, R., and Zeldes, H. (1970). *J. Chem. Phys.* **53**, 1406.
McGlynn, S. P., Azumi, T., and Kinoshita, M. (1969). "Molecular Spectroscopy of the Triplet State", Prentice Hall, Englewood Cliffs, N.J.
Maruyama, K., and Otsuki, T. (1971). *Bull. Chem. Soc. Jap.* **44**, 2885.
Maruyama, K., Shindo, H., and Maruyama, T. (1971a). *Bull. Chem. Soc. Jap.* **44**, 585.

Maruyama, K., Otsuki, T., Shindo, H., and Maruyama, T. (1971b). *Bull. Chem. Soc. Jap.* **44**, 2000.

Maruyama, K., Shindo, H., Otsuki, T., and Maruyama, T. (1971c). *Bull. Chem. Soc. Jap.* **44**, 2756.

Maruyama, K., Otsuki, T., and Takuwa, A. (1972). *Chem. Lett.* **131**.

Méchin, B., and Naulet, N. (1972). *J. Organometal. Chem.* **39**, 229.

Mill, T., and Stringham, R. S. (1969). *Tetrahedron Lett.* 1853.

Mitchell, T. N. (1972). *Tetrahedron Lett.* 2281.

Morris, D. G. (1969). *Chem. Comm.* 1345.

Morris, D. G. (1971a). *Chem. Comm.* 221.

Morris, D. G. (1971b). *Ind. Chim. Belge* **36**, 1060.

Morris, J. I., Morrison, R. C., Smith, D. W., and Garst, J. F. (1972). *J. Am. Chem. Soc.* **94**, 2406.

Müller, K. (1972). *Chem. Comm.* 45.

Müller, K., and Closs, G. L. (1972). *J. Am. Chem. Soc.* **94**, 1002.

Murrell, J. N., and Teixeira-Dias, J. J. C. (1970). *Mol. Phys.* **19**, 521.

Neta, P., Fessenden, R. W., and Schuler, R. H. (1971). *J. Phys. Chem.* **75**, 1654.

Noyes, R. M. (1954). *J. Chem. Phys.* **22**, 1349.

Noyes, R. M. (1955). *J. Am. Chem. Soc.* **77**, 2042.

Noyes, R. M. (1956). *J. Am. Chem. Soc.* **78**, 5486.

Noyes, R. M. (1961). "Progress in Reaction Kinetics" (G. Porter, ed.), Vol. 1, Pergamon, Oxford, p. 129.

Ostermann, G., and Schöllkopf, U. (1970). *Liebigs Ann. Chem.* **737**, 170.

Pedersen, J. B., and Freed, J. H. (1972). *J. Chem. Phys.* **57**, 1004.

Porter, N. A., Marnett, L. J., Lochmüller, C. H., Closs, G. L., and Shobataki, M. (1972). *J. Am. Chem. Soc.* **94**, 3664.

Rakshys, J. W. (1970). *Chem. Comm.* 578.

Rakshys, J. W. (1971). *Tetrahedron Lett.* 4745.

Rieker, A. (1971). *Ind. Chim. Belge* **36**, 1078.

Rieker, A., Niederer, P., and Leibfritz, D. (1969). *Tetrahedron Lett.* 4287.

Rieker, A., Niederer, P., and Stegmann, H. B. (1971). *Tetrahedron Lett.* 3873.

Rosenfeld, S. M., Lawler, R. G., and Ward, H. R. (1972). *J. Am. Chem. Soc.* **94**, 9255.

Roth, H. D. (1971a). *J. Am. Chem. Soc.* **93**, 1527.

Roth, H. D. (1971b). *J. Am. Chem. Soc.* **93**, 4935.

Roth, H. D. (1971c). *Ind. Chim. Belge* **36**, 1068.

Roth, H. D. (1972a). *J. Am. Chem. Soc.* **94**, 1400.

Roth, H. D. (1972b). *J. Am. Chem. Soc.* **94**, 1761.

Roth, H. D., and Lamola, A. A. (1972). *J. Am. Chem. Soc.* **94**, 1013.

Rüchardt, C., Merz, E., Freudenberg, B., Opgenorth, H.-J., Tan, C. C., and Werner, R. (1970). *In* "Essays on Free Radical Chemistry", Chemical Society Special Publication, No. 24, p. 51.

Rykov, S. V., and Buchachenko, A. L. (1969). *Dokl. Akad. Nauk SSSR* **185**, 870.

Rykov, S. V., and Sholle, V. D. (1971a). *Izv. Akad. Nauk SSSR, Ser. Khim.* 2238.

Rykov, S. V., and Sholle, V. D. (1971b). *Izv. Akad. Nauk SSSR, Ser. Khim.* 2351.

Rykov, S. V., Buchachenko, A. L., Dodonov, V. A., Kessenikh, A. V., and Razuvaev, G. A. (1969a). *Dokl. Akad. Nauk SSSR* **189**, 341.

Rykov, S. V., Buchachenko, A. L., and Baldin, V. I. (1969b). *Zh. Strukt. Khim.* **10**, 928.

Rykov, S. V., Buchachenko, A. L., and Kessenikh, A. V. (1970a). *Spectrosc. Lett.* **3**, 55.

Rykov, S. V., Buchachenko, A. L., and Kessenikh, A. V. (1970b). *Kinet. Katal.* **11**, 549.

Schöllkopf, U. (1970). *Angew. Chem.* **82**, 795; *Int. Ed.* **9**, 763.

Schöllkopf, U. (1971). *Ind. Chim. Belge* **36**, 1057.

Schöllkopf, U., Ostermann, G., and Schossig, J. (1969a). *Tetrahedron Lett.* 2619.

Schöllkopf, U., Ludwig, U., Ostermann, G., and Patsch, M. (1969b). *Tetrahedron Lett.* 3415.

Schöllkopf, U., Schossig, J., and Ostermann, G. (1970). *Liebigs Ann. Chem.* **737**, 158.

Schulman, E. M., Bertrand, R. D., Grant, D. M., Lepley, A. R., and Walling, C. (1972). *J. Am. Chem. Soc.* **94**, 5972.

Shindo, H., Maruyama, K., Otsuki, T., and Maruyama, T. (1971). *Bull. Chem. Soc. Jap.* **44**, 2789.

Shine, H. J. (1967). "Aromatic Rearrangements", Elsevier, Amsterdam, London and New York, p. 284.

Simons, J. P. (1971). "Spectroscopy and Photochemistry", Wiley, New York, Ch. 3.

Slonim, I. Y., Urman, Y. G., and Konovalov, A. G. (1970). *Dokl. Akad. Nauk SSSR* **195**, 1153.

Smaller, B., Avery, E. C., and Remko, J. R. (1971). *J. Chem. Phys.* **55**, 2414.

Smaller, B., Remko, J. R., and Avery, E. C. (1968). *J. Chem. Phys.* **48**, 5174.

Strausz, O. P., Lown, J. W., and Gunning, H. E. (1972). *In* "Comprehensive Chemical Kinetics", (C. H. Bamford and C. F. H. Tipper, eds.), Vol. 5, Elsevier, Amsterdam, London and New York, Ch. 5.

Tomkiewicz, M., and Klein, M. P. (1972). *Rev. Sci. Inst.* **43**, 1206.

Tomkiewicz, M., Groen, A., and Cocivera, M. (1971). *Chem. Phys. Lett.* **10**, 39.

Tomkiewicz, M., Groen, A., and Cocivera, M. (1972). *J. Chem. Phys.* **56**, 5850.

Tsolis, A., Mylonakis, S. G., Nieh, M. T., and Seltzer, S. (1972). *J. Am. Chem. Soc.* **94**, 829.

Urry, W. H., and Eiszner, J. R. (1952). *J. Am. Chem. Soc.* **74**, 5822.

Urry, W. H., Eiszner, J. R., and Wilt, J. W. (1957). *J. Am. Chem. Soc.* **79**, 918.

Veeman, W. S., and van der Waals, J. H. (1970). *Chem. Phys. Lett.* **7**, 65.

Walling, C., and Gibian, M. V. (1965). *J. Am. Chem. Soc.* **87**, 4313.

Walling, C., and Lepley, A. R. (1971). *J. Am. Chem. Soc.* **93**, 546.

Walling, C., and Lepley, A. R. (1972). *J. Am. Chem. Soc.* **94**, 2007.

Walling, C., Waits, H. P., Milovanovic, J., and Pappiaonnou, C. G. (1970). *J. Am. Chem. Soc.* **92**, 4927.

Ward, H. R. (1971). *Ind. Chim. Belge* **36**, 1085.

Ward, H. R. (1972). *Accounts Chem. Res.* **5**, 18.

Ward, H. R., and Lawler, R. G. (1967). *J. Am. Chem. Soc.* **89**, 5518.

Ward, H. R., Lawler, R. G., and Loken, H. Y. (1968). *J. Am. Chem. Soc.* **90**, 7359.

Ward, H. R., Lawler, R. G., and Cooper, R. A. (1969a). *Tetrahedron Lett.* 527.

Ward, H. R., Lawler, R. G., and Cooper, R. A. (1969b). *J. Am. Chem. Soc.* **91**, 746.

Ward, H. R., Lawler, R. G., Loken, H. Y., and Cooper, R. A. (1969c). *J. Am. Chem. Soc.* **91**, 4928.

Ward, H. R., Lawler, R. G., and Marzilli, T. A. (1970). *Tetrahedron Lett.* 521.

Wigfield, D. C., Feiner, S., and Taymaz, K. (1972). *Tetrahedron Lett.* 891.

Williams, G. H. (1970). *In* "Essays on Free Radical Chemistry", Chemical Society Special Publication, No. 24, p. 25.

Woodward, R. B., and Hoffmann, R. (1970). "The Conservation of Orbital Symmetry", Verlag Chemie, Weinheim.

Yang, N. C., and Feit, E. D. (1968). *J. Am. Chem. Soc.* **90**, 504.

Yang, N. C., Feit, E. D., Hui, M. M., Turro, N. J., and Dalton, J. C. (1970). *J. Am. Chem. Soc.* **92**, 6974.

Zeldes, H., and Livingston, R. (1970). *J. Phys. Chem.* **74**, 3336.

THE PHOTOCHEMISTRY OF CARBONIUM IONS

P. W. CABELL-WHITING AND H. HOGEVEEN

Department of Organic Chemistry, The University, Zernikelaan, Groningen, The Netherlands

I. Introduction.	129
II. Valence Isomerization Reactions	130
A. Non-benzenoid Aromatics .	130
B. Alkylbenzenium Ions	133
C. Protonated Cyclohexadienones .	137
D. Heteroaromatics	139
E. Protonated Cycloheptadienones	142
III. Electron Transfer and Coupling Reactions	145
A. Cyclopropenyl Cations	145
B. Triphenylmethyl (Trityl) Cation.	145
IV. Conclusions .	150
References .	151

I. INTRODUCTION

IN past decades, the vast and once uncharted territory comprised of carbonium ion chemistry has been researched and reorganized to yield quantified bundles of information of both theoretical and practical use to the chemist. This process was evolved in large measure as an attempt to catalogue and characterize the various types of reactions these species undergo. Numerous books and reviews have appeared in the literature dealing with the method of formation and various types of carbonium ions, their thermodynamic aspects, electronic, vibrational, and nuclear magnetic resonance spectra, electrolytic conductivity, bridgehead reactivity, and their classical or non-classical structure, among other things. Carbonium ion reactions have been investigated with respect to the influence of stereochemical factors, solvent effects, and lifetimes of the transient intermediates. A new vocabulary has been devised, in fact, in order to enable chemists to discuss these species in a meaningful way.

However, despite the detailed attention which has been given to many facets of carbonium ion chemistry, there has been, until recently, very little information available regarding their photoreactions. It is specifically through the use of the "super acid" solvents, e.g. $HF-SbF_5$, $FHSO_3-SbF_5$, $HF-BF_3$, etc., that a vast number of stable, long-lived carbonium ions have now become available for study. Many otherwise transient intermediates are found to have appreciable lifetimes in very strong acid at low temperatures, and can, thus, be characterized and their photoreactions studied with a minimum of experimental difficulty. It is our purpose, therefore, to present in this review what is known with respect to the photochemistry of such charged species. In undertaking to delineate the reactions in question, we have chosen to examine systems in light of the types of reactions they undergo and the sorts of intermediates they involve.

The first example cited in the literature of a cationic photolysis in which products were isolated and identified, was the observation that trityl perchlorate, exposed to ordinary overhead fluorescent lighting for a period of about two weeks, resulted in the formation of 9-phenylfluorene as a major product, along with several minor ones which were not identified (Dauben, 1960). This was not considered of such importance as to merit further investigation, and it was some several years later that chemists first began to study directly the photochemistry of ionic carbon species, i.e. carbonium ions, and to a lesser extent, carbanions, from the standpoint of product development and mechanism. In this review we shall only deal with the former class, and we shall begin by investigating the most commonly observed reaction which these charged species undergo, namely valence isomerizations, i.e. reactions in which only carbon–carbon and not carbon–hydrogen bonds are involved.

II. Valence Isomerization Reactions

A. *Non-benzenoid Aromatics*

The first carbocationic photolysis to be investigated in detail was that of the tropylium ion (1) (van Tamelen *et al.*, 1968, 1971), in which generation of the [3,2,0] valence bond isomer (the "Dewar tropylium ion"), (2) was the dominant reaction. When irradiated for 10 minutes in 5% aqueous sulfuric acid, two major products were formed in a total

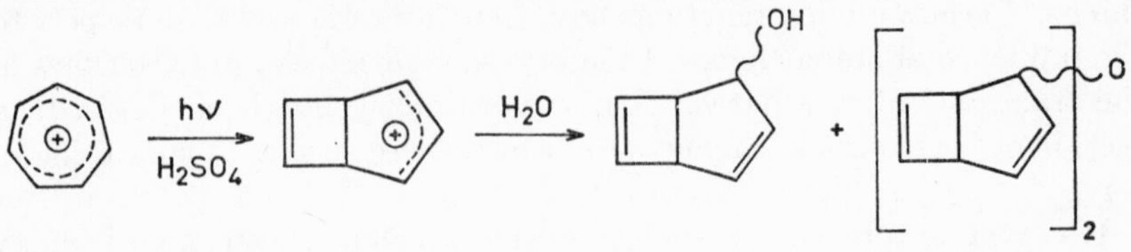

(1)   (2)   (3)   (4)

yield of 58%: bicyclo[3,2,0]hepta-3,6-dien-2-ol (3) and the derived ether (4), both arising from the allylic carbonium ion (2) which readily captures solvent water.

That the photoreactive species is the carbonium ion and not the corresponding alcohol is clearly indicated by the relative concentrations of the two species present. The calculated equilibrium constant for 5% aqueous sulfuric acid implies an alcohol content of $2.7 \times 10^{-4}\%$, much too low to account for any detectable photoreaction from this covalent species. In addition, when the tropylium salt is irradiated in the absence of acid, neither 3 nor 4 is detected as a product, but rather ditropyl (5) and its photoisomer (6) are observed.

(5)   (6)

When irradiated in stronger acid ($FHSO_3$) at $-60°$, the tropylium ion is seen to isomerize cleanly to the norbornadien-7-yl cation (7) (Childs and Taguchi, 1970; Hogeveen and Gaasbeek, 1970). The ion 7 has been shown to be photochemically stable (Cabell and Hogeveen, 1972). However, it is known to revert to the tropylium ion thermally at temperatures above 45° (Lustgarten et al., 1967).

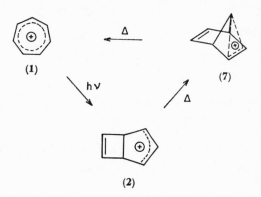

The most reasonable mechanism for this transformation, in accord with that suggested by van Tamelen et al. for the dilute acid photolysis, is initial photoisomerization to 2 followed in this case by thermal conversion of the "Dewar tropylium ion" to 7. The isomerization of 2 to 7 has been reported to be very rapid at temperatures below $-60°$, and it has been shown, in addition that, in nucleophilic solvents, capture of 2 competes very efficiently with isomerization (Lustgarten et al., 1967).

Methyltropylium ion photolyzed at $-85°$ in $FHSO_3$ afforded the 2-methylnorbornadienyl cation as a major product. Phenyltropylium ion (8), however, displays a significantly different mode of reaction from the parent ion (van Tamelen *et al.*, 1971). When irradiated in aqueous sulfuric acid under anaerobic conditions, the phenyl-substituted compound yields no products. In the presence of oxygen, however, biphenyl (9) is formed, postulated to occur as the result of action by hydrogen peroxide on the ion (Jutz and Voithenleiter, 1964).

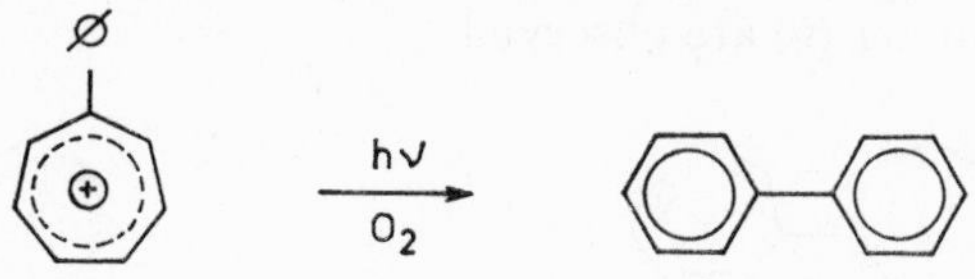

(8)   (9)

Photolysis in acetonitrile gives *ortho*- (10) and *para*-phenylbenzaldehyde (11), as well as *cis*- (12) and *trans*-diphenylstilbene (13). The mechanism for product formation is not clear, but the norcaradiene valence isomer is thought to be the most likely precursor, arising either via a secondary

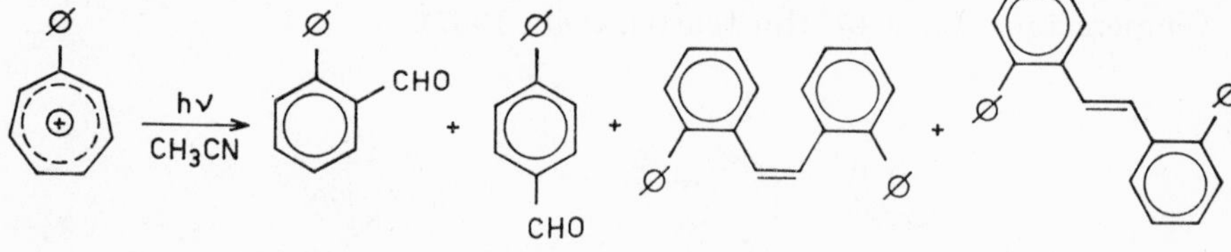

(8)   (10)   (11)   (12)   (13)

thermal reaction from vibrationally excited phenyltropylium ion or from photo-excitation of norcaradiene which is in thermal equilibrium with 8. Irradiation of the homotropylium ion (14) at $-70°$ in fluorosulfonic acid results in the formation of a single stable product (Hogeveen and Gaasbeek, 1970). However, no definite structure has been assigned to this product as yet.

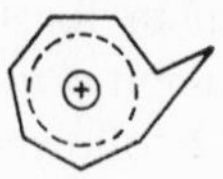

(14)

The azulenium ion (15), while possessing the characteristic tropylium moiety, by merit of conjugation of the aromatic nucleus with a double bond, might be expected to undergo significantly different photochemical reactions from 1 (van Tamelen *et al.*, 1968, 1971). In fact, when 15 is irradiated in 50% aqueous sulfuric acid, two products, **16** and **17**, are isolated in addition to polymer.

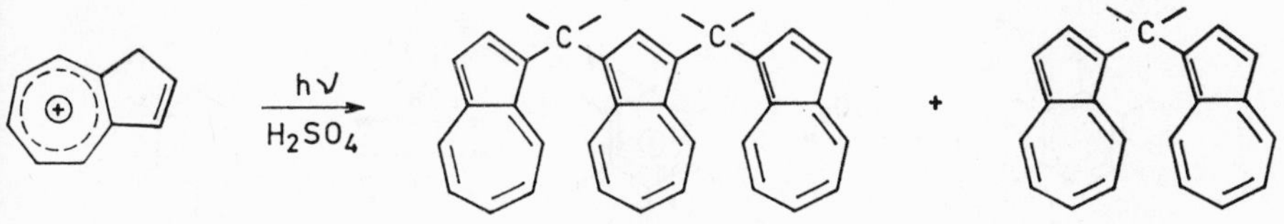

(15)   (16)   (17)

The mechanism for this photoreaction has not been elucidated, although it has been ascertained that these products are not observed in the dark reaction with acetone under the same conditions as used in the photolysis. The source of the isopropylidene unit is undetermined, but since all extraneous organic sources were excluded in the experiment, it seems clear that it originates from the azulene system itself, presumably from the five-membered ring.

B. *Alkylbenzenium Ions*

A valence isomerization reaction similar to that encountered with the tropylium ion has been observed when a variety of substituted benzenes are photolyzed in strong acid. The pentamethylbenzenium ion (**18**) photolyzed at $-78°$ to give a single product, the pentamethylbicyclo-[3,1,0]hexenyl cation (**19**), in excess of 80% conversion (Childs *et al.*, 1968).

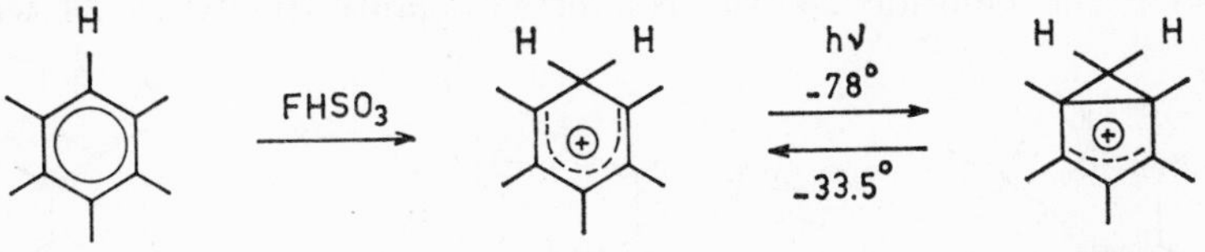

(18)   (19)

It is of substantial interest to note that, as the temperature of the reaction mixture is increased to $-33.5°$, ion **19** is converted quantitatively back to **18**. At that temperature the first-order rate constant for the reversion has been calculated to be 8.0×10^{-4} sec^{-1}, which corresponds to a free enthalpy barrier ($\Delta G^{\ddagger}$) of 17.4 kcal/mol.

The hexamethyl- and heptamethyl-benzenium ions (**20** and **21**) irradiated under the same conditions gave the corresponding bicyclic [3,1,0] products (**22** and **23**) (Childs and Winstein, 1968). Of the two conceivable stereoisomers which can arise from **20**, only one is observed.

This has been identified as the *endo* isomer, i.e. the one in which the C_6 methyl group is "inside" (Koptyug *et al.*, 1969). As in the case of **19**, when **22** is warmed to $-34°$, a clean first-order reversion to the hexa-

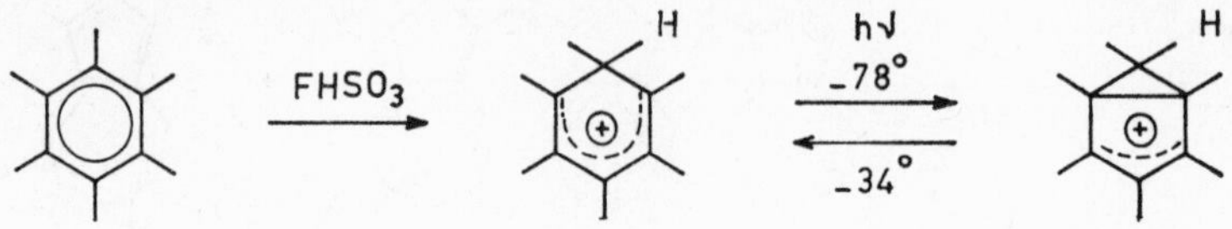

methylbenzenium ion occurs with a rate constant of $1 \cdot 15 \times 10^{-3}$ sec^{-1}, the $\Delta G^{\ddagger}$ value being $17 \cdot 1$ kcal/mole. It was argued that orbital symmetry predicts a conrotatory mode of ring opening for the ground state reaction, clearly precluded in this ring system (Woodward and Hoffmann, 1970). Thus, the conversion, if it were concerted, must proceed in a "forbidden" disrotatory fashion, indicating that electronic considerations are overruled by thermodynamic arguments in product determination. Independent experiments, however, showed that in SbF$_5$—SO$_2$ at $-60°$ the chloride **24** was converted quantitatively to **20** within

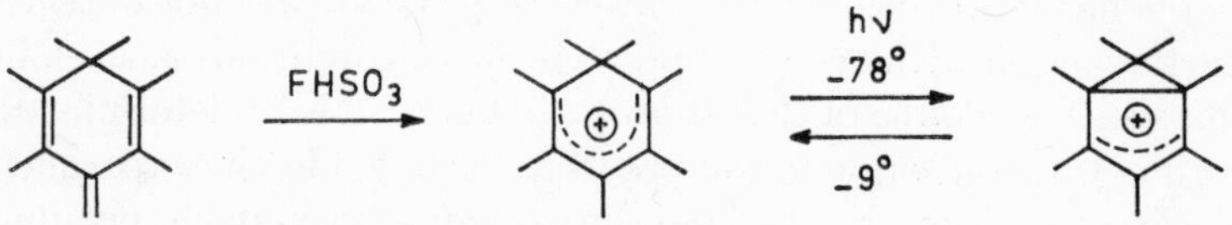

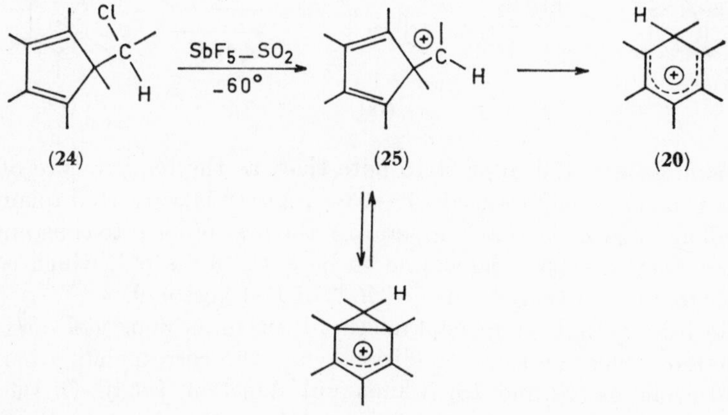

8 minutes (Hogeveen and Kwant, 1972). This suggested the strong possibility that, in contrast to the symmetry-allowed concerted photochemical ring closure of 20 to 22, the reverse thermal reaction is a stepwise one, involving initial cleavage of one cyclopropyl bond (β-cleavage) to yield the intermediate 25. Since the half-life of 25 at $-60°$ is certainly less than one minute, it is not surprising that its presence was not detected at $-34°$.

Thermal reversion of 23 to 21 occurs at $-9°$, $k = 2\cdot2 \times 10^{-4}$ sec^{-1}; $\Delta G^{+} = 19\cdot8$ kcal/mol. The spectrum of 23 is temperature-dependent and when generated in FHSO$_3$—SO$_2$ClF could be scanned from $-110°$ to $-9°$. N.m.r. line broadening experiments indicate the occurrence of a five-fold degenerate scrambling which is a result of migration of the cyclopropyl carbon atom C$_6$ around the cyclopentenyl ring. It can be best viewed as a concerted sigmatropic 1,4 shift, symmetry-allowed in a suprafacial manner with inversion of C$_6$ (Woodward and Hoffmann, 1970).

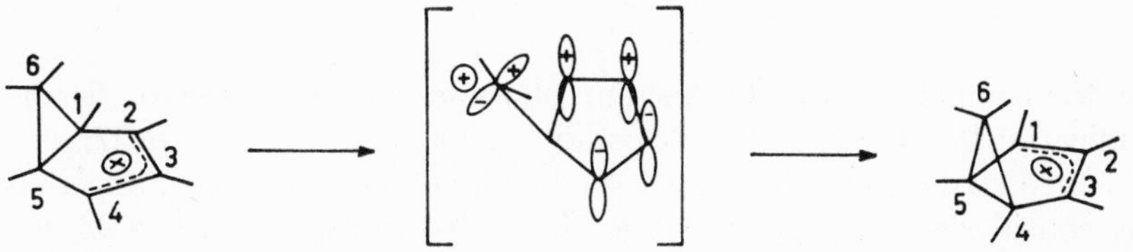

While the pentamethylbicyclo[3,1,0]hexenyl cation (19) and the *endo*-hexamethyl cation (22) exhibit no such behavior observable on the n.m.r. time scale, the independently synthesized *exo*-hexamethylbicyclo-[3,1,0]hexenyl isomer does undergo the same phenomenon. Thus, this type of degenerate rearrangement seems to be very sensitive to the degree of substitution of the C$_6$ atom, implying that the migrating atom bears a considerable positive charge during the transition state, and, in addition, to the steric arrangement of the substituents at this atom.

Investigation of the photochemistry of protonated durene offers conclusive evidence that the mechanism for isomerization of alkylbenzenium ions to their bicyclic counterparts is, indeed, a symmetry-allowed disrotatory closure of the pentadienyl cation, rather than a $[\sigma 2_a + \pi 2_a]$ cycloaddition reaction, which has been postulated to account for many of the photoreactions of cyclohexadienones and cyclohexenones (Woodward and Hoffmann, 1970). When the tetramethyl benzenium ion (26) is irradiated in FHSO$_3$ at $-90°$, the bicyclo[3,1,0]hexenyl cation (27) is formed exclusively (Childs and Parrington, 1970). If photoisomerization had occurred via a $[\sigma 2_a + \pi 2_a]$ cycloaddition, the expected product would have been **28**. Since no **28** is formed during photolysis this pathway can be ruled out. As the reaction mixture is heated to $-31°$, clean conversion back to **26** is observed ($k = 1 \cdot 46 \times 10^{-3}$ s^{-1}; $\Delta G^{\ddagger} = 17 \cdot 1$ kcal/mol) with no **28** detected; and less than 5% of its expected thermal progeny, protonated isodurene (**29**). If **29** is irradiated

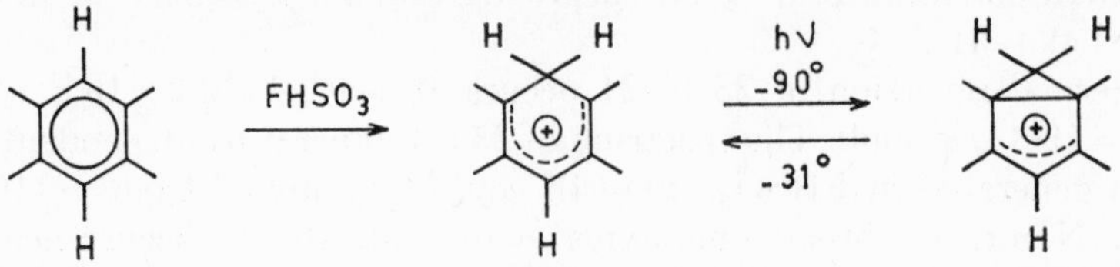

(26)   (27)   (28)   (29)

under comparable conditions, no photoproduct is detected, further indicating that a $[\sigma 2_a + \pi 2_a]$ cycloaddition does not occur to give **27**. That the predicted disrotatory closure product is not observed can be explained by considering that, given the methyl substitution pattern of **28**, it is not unreasonable to expect this compound to be thermally unstable, even at $-90°$.

Once again we see that in the absence of methyl substitution on the C$_6$ carbon, cyclopropyl migration cannot be observed on the n.m.r. time scale. It is interesting to note that, although degenerate scrambling in these systems is dependent upon the degree of substitution and stereochemistry at C$_6$, the magnitude of the free enthalpy barrier to ring opening of the bicyclohexenyl cations is not.

When the pentamethylbenzyl cation (**30**) is irradiated at $-80°$ in FHSO$_3$ for 4 hours, one photoproduct is observed in 50% yield (Cabell and Hogeveen, 1972). Its n.m.r. spectrum is identical to that observed for the hexamethylbicyclo[3,1,0]hexenyl cation (**22**) generated in the

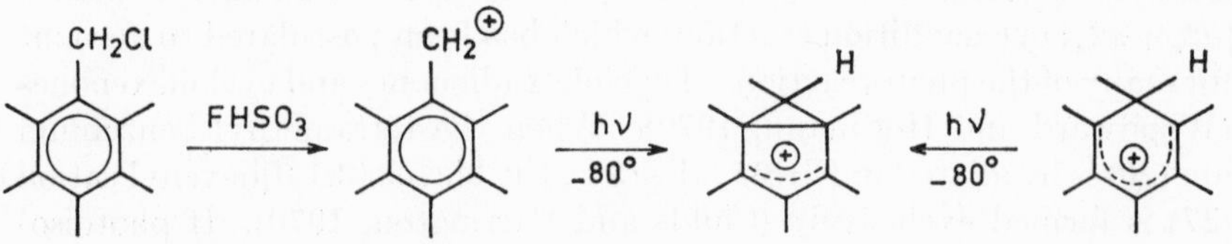

(30)   (22)   (20)

photolysis of the hexamethylbenzenium ion (**20**). Although a mechanism has not been established for this reaction, it is possible that the reaction involves photoreduction of **30** to yield **20** followed by ring closure to yield **22**.

Benzene itself, when photolyzed in acetic and aqueous phosphoric acids, has been reported to yield photo adducts **31** and **32** respectively (Farenhorst and Bickel, 1966). Protonated prefulvene (**33**) is suggested

$$hv \atop \text{HOAc or } H_3PO_4(aq)$$

(**31**) : X = OAc
(**32**) : X = OH

as an intermediate, which would arise from irradiation of σ-protonated benzene (Bryce-Smith *et al.*, 1967). Nucleophilic attack on the less hindered side of the carbonium ion would account for the observed stereochemistry of the products.

(**33**)

C. *Protonated Cyclohexadienones*

It has been found that cyclohexa-2,4- and -2,5-dienones undergo a light-induced valence isomerization reaction in strong acid analogous to the alkylbenzenes, to yield 2-hydroxy-bicyclo[3,1,0]hexenyl cations. The hydroxybenzenium ion (**34**), for example, underwent a clean photo-isomerization to **35** at temperatures below $-60°$ (Parrington and Childs, 1970). Cation **35** was also produced upon similar irradiation of **36**.

The lack of any difference in the rate of isomerization between fluoro-sulfonic acid solutions of **34** which had been thoroughly degassed, and those which were saturated with oxygen, suggests that the reaction does not proceed via a triplet mechanism. In fluorosulfonic acid no unproton-ated acid is detected, ruling out the possibility of n,π* excitation. Thus, there is little doubt in this case that it is the π,π* singlet state which is the reactive species. Experiments carried out with a variety of methyl-substituted protonated cyclohexadienones have likewise ruled out the possibility of photochemically induced alkyl migrations during the first step of the reaction. Rather initial valence isomerization of **36** to **38** is suggested, followed by a thermal cyclopropyl migration to give **35**. Cation **37** is known to be thermally accessible from and to revert to **35** by what is formally a suprafacial 1,4-sigmatropic shift (Hart *et al.*, 1969). The same mechanism can also be seen to hold for the transformation of **36** to **35**, with the cation **38** as the intermediate formed photochemically in the first step. We observe in the case of protonated cyclohexadienones that cyclopropyl migration appears to be, as with the alkylbenzenes, greatly influenced by the charge-stabilizing ability of the substituents on the C_6 atom.

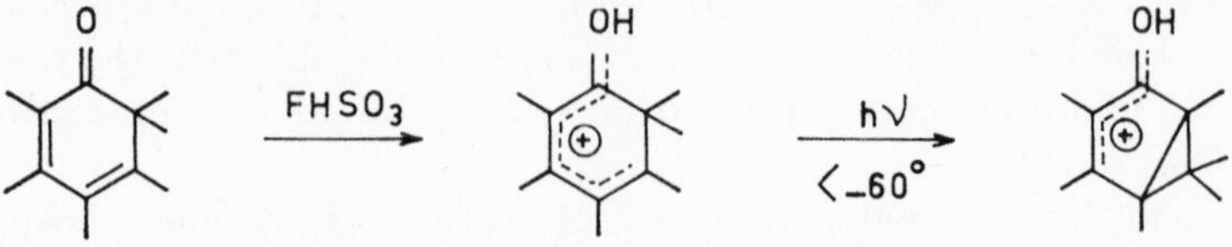

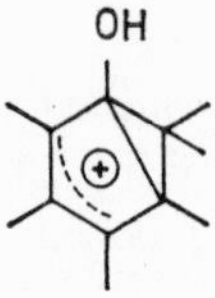

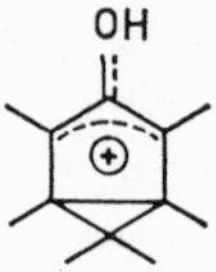

When dienones **39** and **40** are photolyzed in sulfuric acid they both rearrange to the same product, 2-methyl-5-hydroxybenzaldehyde (**41**) (Filipescu and Pavlik, 1970). The mechanism for this photorearrangement is consistent with that of the protonated cyclohexadienones already discussed, i.e., disrotatory closure to afford the intermediate bicyclic cations **42** and **43**. In this case it is conceivable that the electron-withdrawing effect of the dichloromethyl group forces the subsequent thermal cyclopropyl migration entirely in the direction of the most stable cation **44** to yield the observed product.

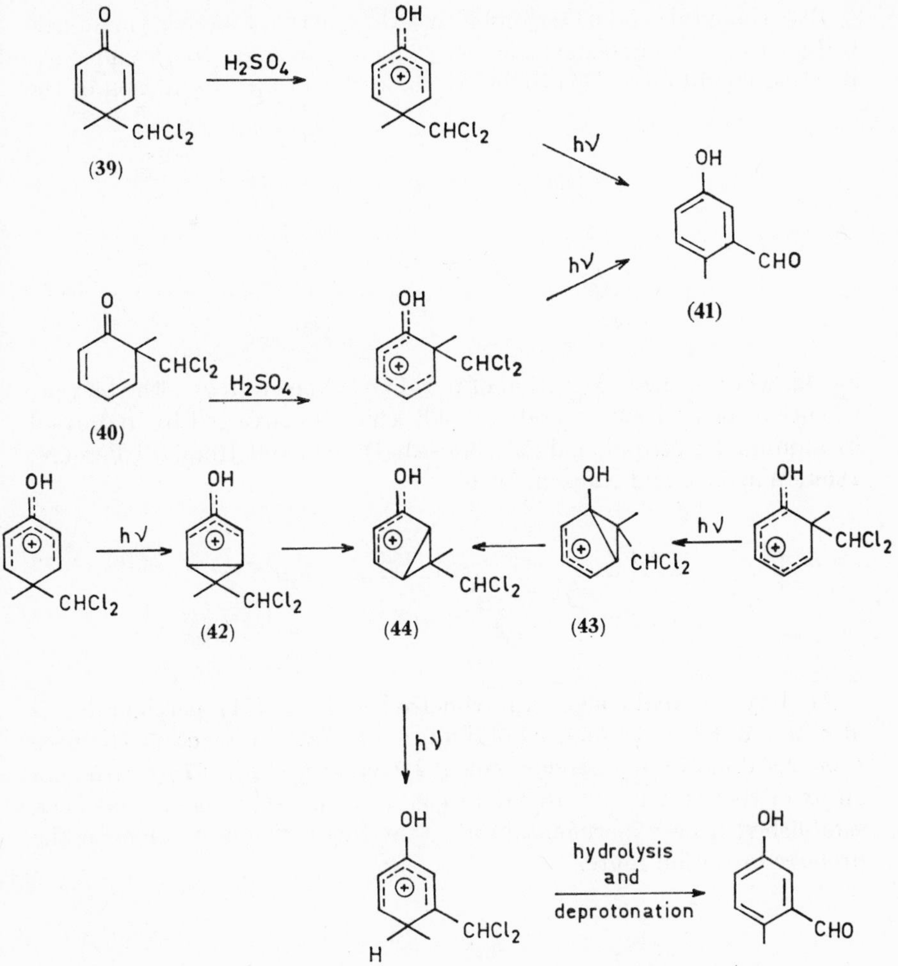

D. *Heteroaromatics*

Little study has been undertaken on the photoreactions of hetero-aromatic cations; however, the photohydration of the methylpyridinium ion (45) to yield 6-methylazabicyclo[3,1,0]hex-3-en-2-*exo*-ol (46) has been reported (Kaplan *et al.*, 1972).

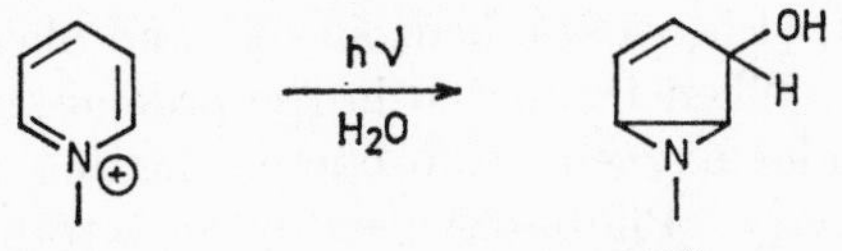

Experimental evidence suggests that the reaction proceeds via isomerization to an azoniabenzvalene (47) which rearranges to 48 under the reaction conditions. This is clearly the result of π,π^* excitation of the

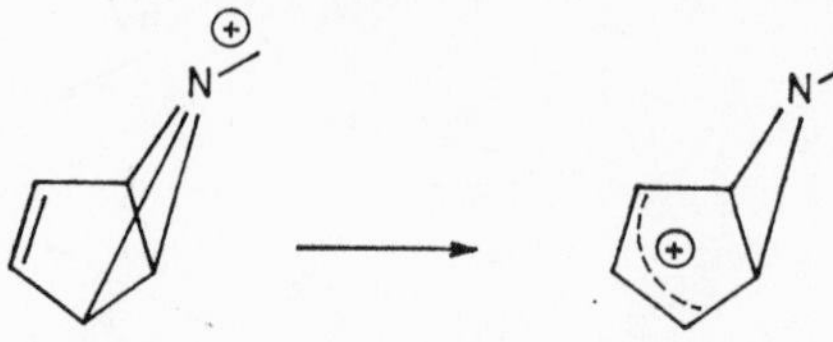

(47)   (48)

ion 45, whereas n,π^* excitation of pyridine is known to give the bicyclic valence isomer "Dewar pyridine" (49) which is converted by hydration to 5-amino-2,4-pentadienal (50) (Joussot-Dubien and Houdard-Pereyre, 1969; Wilzbach and Rausch, 1970).

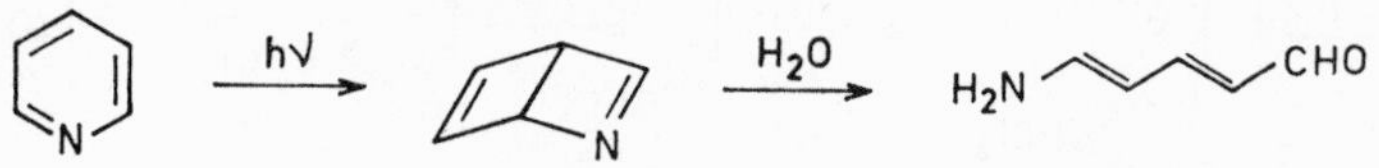

(49)   (50)

An oxygen analogue, 2,4,6-trimethylpyrilium (51) perchlorate, is also found to undergo a photoinduced hydration reaction, to give trans-2,4-dimethyl-5-oxohex-2-enal (52) (Barltrop et al., 1972). Although an oxoniabenzvalene intermediate has been proposed, it has not been established from experimental data that this is the only, or even the preferred reaction route.

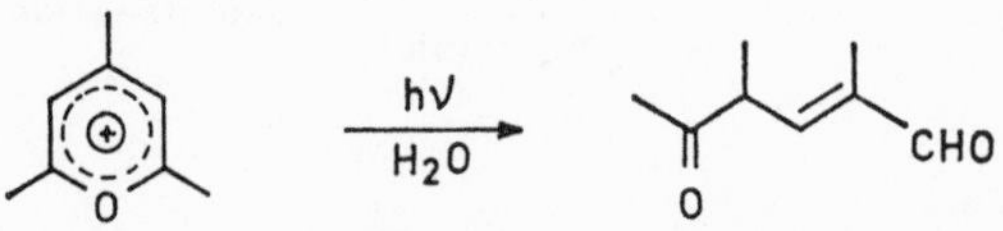

(51)   (52)

The 6π non-benzenoid aromatic pyrilium system has also been observed to undergo a photooxygenation reaction, along with its sulfur analog (Yoshida et al., 1971). The 2,4,6-triphenylpyrilium (53) and thiopyrilium (54) cations, irradiated in the presence of oxygen, yielded benzaldehyde, methyl benzoate, benzoic acid and phenol or thiophenol as major products. When irradiated in the presence of methylene blue as a sensitizer, under oxygen atmosphere, starting material was recovered quantitatively. Photo-oxygenation of tetramethylethylene in

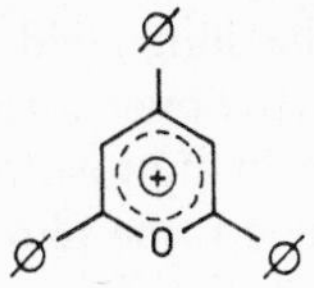

(53) 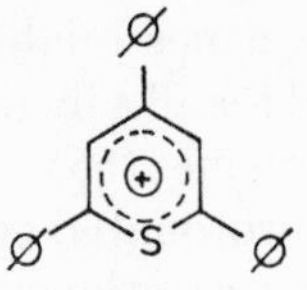   (54)

the presence of methylene blue and either **53** or **54**, demonstrated that the cations do not act as quenchers under such conditions. Thus, it appears that the reaction is one in which ground state oxygen reacts with a triplet excited state carbonium ion, and not the reaction of a ground state carbonium ion with singlet oxygen. This is reasonable in the light of the fact that singlet oxygen is generally considered to be electrophilic (Foote, 1968).

The probable reaction mechanism involves formation of the peroxide **55**, which then fragments to give the observed products.

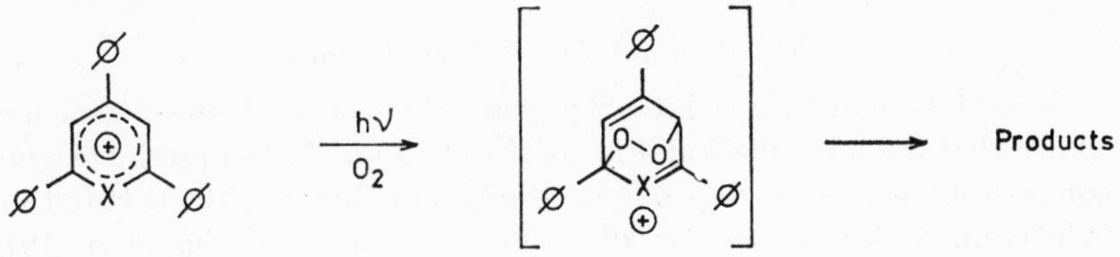

(55)

(53) : X = O
(54) : X = S

Since thiophenes are known to undergo a variety of reactions in a manner analogous to benzene, it is not unreasonable to suppose that

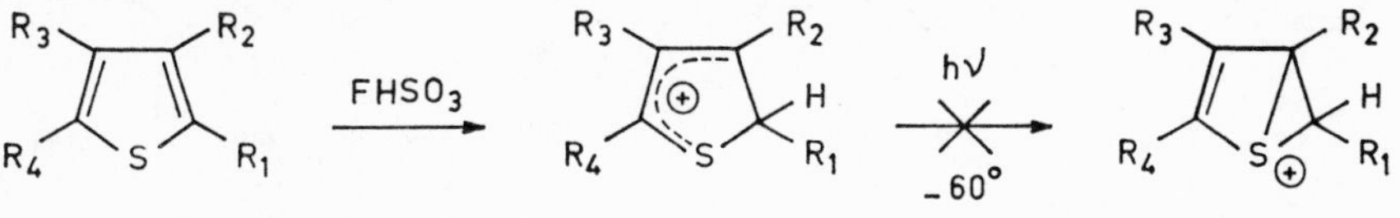

(56)   a : $R_1 = R_2 = R_3 = R_4 = H$
       b : $R_4 = (CH_3)_3C$; $R_1 = R_2 = R_3 = H$
       c : $R_1 = R_4 = (CH_3)_3C$; $R_2 = R_3 = H$
       d : $R_4 = CH_3$; $R_1 = R_2 = R_3 = H$
       e : $R_1 = R_4 = CH_3$; $R_2 = R_3 = H$
       f : $R_1 = R_2 = R_3 = R_4 = CH_3$
       g : $R_4 = C_6H_5$; $R_1 = R_2 = R_3 = H$

protonated thiophenes might, upon irradiation, yield products similar to those observed for alkylbenzenium ions. However, although a variety of substituted thiophenes (56) were irradiated in fluorosulfonic acid at − 60° for 5 hours, the protonated species were seen to be photostable, and the expected bridging reaction did not take place (Hogeveen et al., 1972). By contrast, irradiation of S-methyltetramethylthiophenium ion (57) in CH$_3$OSO$_2$F for 5 hours at room temperature affords one product in ~ 80% yield. The structure of this photoproduct has been found to be the 2,2,3,4,5-pentamethylthiophenium ion.

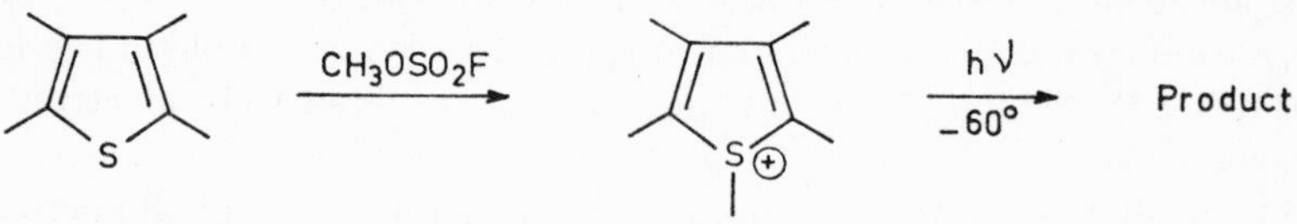

(57)

E. Protonated Cycloheptadienones

Several investigations into the photochemistry of eucarvone have shown that upon irradiation it isomerizes to a mixture of products whose composition is solvent-dependent (Büchi and Burgess, 1960; Hurst and Whitham, 1963; Shuster et al., 1964; Shuster and Sussman, 1970; Takino and Hart, 1970; Hart and Takino, 1971). Ionic intermediates have been invoked in the case of polar solvents (Chapman, 1963). Irradiation of protonated eucarvone 58 in fluorosulfonic acid seems to

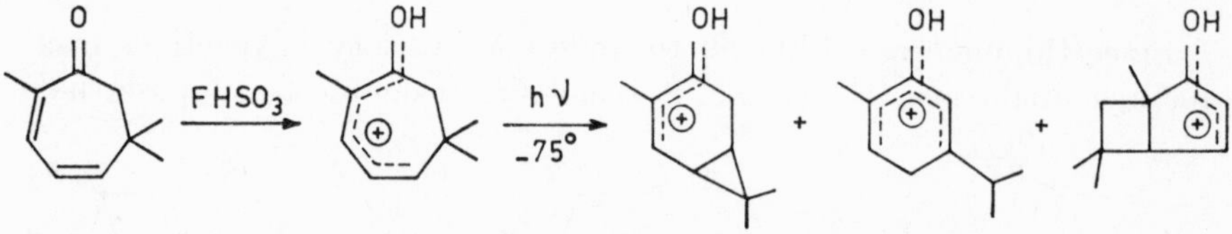

(58)   (59)   (60)   (61)

support the theory of ionic intermediates and results in a photochemical isomerization affording as the major product protonated bicyclo[4,1,0]-heptenone (59) in 73% yield (Hine and Childs, 1971). In previous photolyses the major product has always been the [3,2,0] valence isomer 62, while in FHSO$_3$ no corresponding cation or rearrangement product could be detected.

Formation of 59 and 60 is postulated to occur via an electrocyclic ring opening reaction giving the open chain cation 63 which can then go either

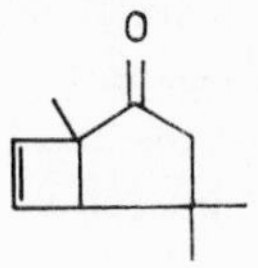

(62)

thermally or photochemically to the cationic species **64**. Such a process has analogy in the photoisomerization of the isoelectronic cyclohexa-1,3-dienes to bicyclo[3,1,0]hexenes (Meinwald and Mazzocchi, 1966).

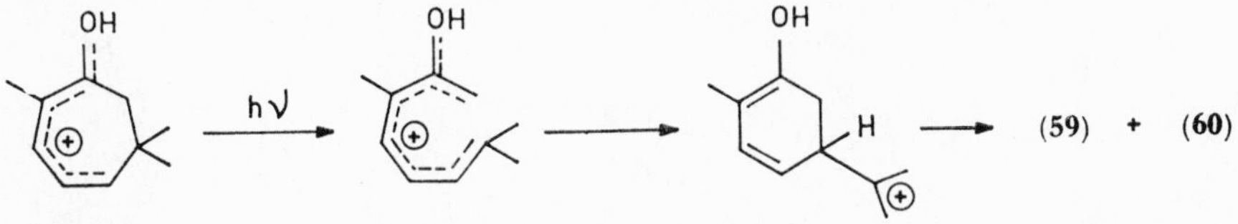

(58)   (63)   (64)

It is likely that **61** arises from the cation **66**, since protonation of the ketone **65** in $FHSO_3$ at $-78°$ gave **66**, which rearranged quantitatively to **61** at $-55°$.

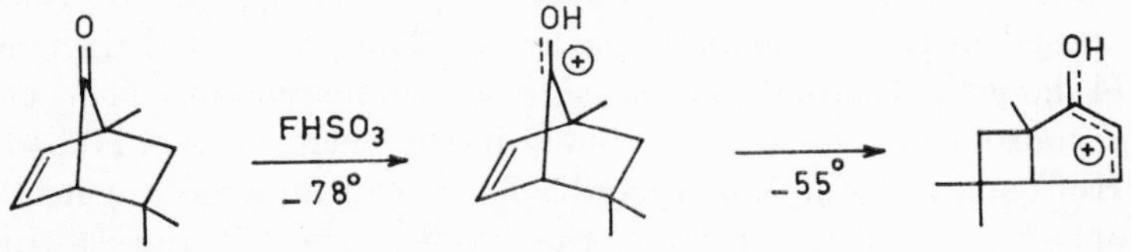

(65)   (66)   (61)

It is interesting to note that the parent dienone **67** and its 2-methyl derivative **68**, when irradiated in strongly acidic media, gave no product which would correspond to the protonated bicyclo[4,1,0]heptenone observed in the photolysis of protonated eucarvone. In $FHSO_3$ at $-60°$

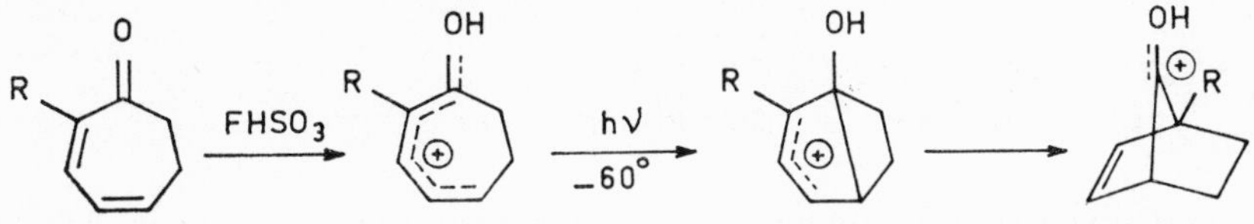

(67) : R = H
(68) : R = CH₃

(71) : R = H
(72) : R = CH₃

(69) : R = H
(70) : R = CH₃

the protonated ketones isomerized to give exclusively protonated norbornen-7-ones **69** and **70** (Hine and Childs, 1972). Formation of these products can be thought of in terms of a photochemically allowed closure between C_1 and C_5 to give the bicyclic cations **71** and **72**, followed by a thermal 1,2-alkyl shift.

The photolysis of protonated 2,6-cycloheptadienone (**73**) and its 5-methyl derivative (**74**) at $-78°$ in $FHSO_3$ yields as a single product in each case the corresponding protonated vinylcyclopentadienones **75** and **76**, which subsequently isomerize thermally to the more stable cations **77** and **78** (Noyori *et al.*, 1972).

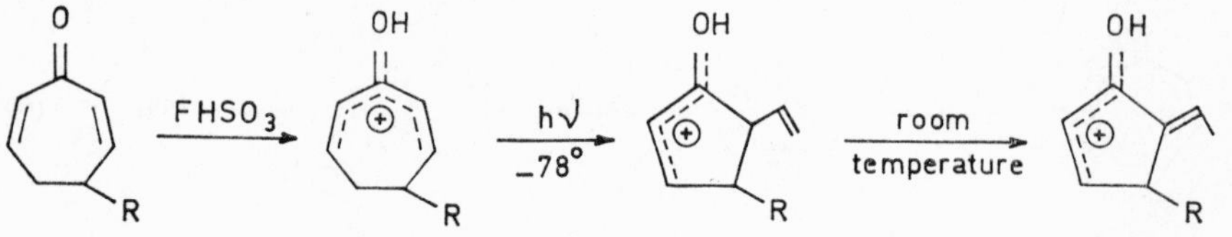

(**73**) : R = H
(**74**) : R = CH_3

(**75**) : R = H
(**76**) : R = CH_3

(**77**) : R = H
(**78**) : R = CH_3

As in the photolysis of protonated eucarvone, an acyclic intermediate is proposed in the mechanistic pathway. The protonated dienones **73** and **74** should be thermally stable, since a symmetry-allowed ring closure in the conrotatory mode is precluded in the cyclic system (Woodward and Hoffmann, 1970). Upon irradiation it can undergo a conrotatory ring opening however, to produce the acyclic cations **79** and **80** which in

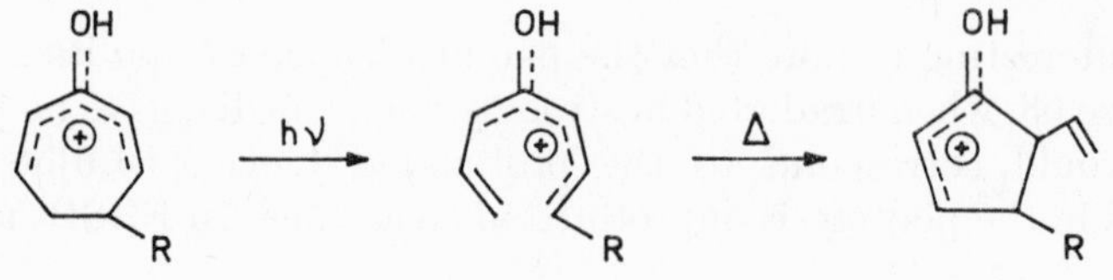

(**73**) : R = H
(**74**) : R = CH_3

(**79**) : R = H
(**80**) : R = CH_3

(**75**) : R = H
(**76**) : R = CH_3

turn may give rise to products via thermal conrotatory cyclization. The facile conversion of the pentadienyl to cyclopentenyl cation is well known, along with the related cyclization of heptatrienyl to vinylcyclopentenyl cations (Sorenson, 1965; Campbell *et al.*, 1969).

III. Electron Transfer and Coupling Reactions

A. *Cyclopropenyl Cations*

The stable triphenylcyclopropenium cation (81) undergoes an electron-transfer reaction when photolyzed in acidic medium (van Tamelen *et al.*, 1968, 1971). Irradiation of 81 for 4 hours in 10% aqueous sulfuric acid resulted in a 49% yield of hexaphenylbenzene (82). The reaction is presumed to proceed by initial charge transfer to produce the cyclopropenyl radical 83, which then couples to give 84. This compound in

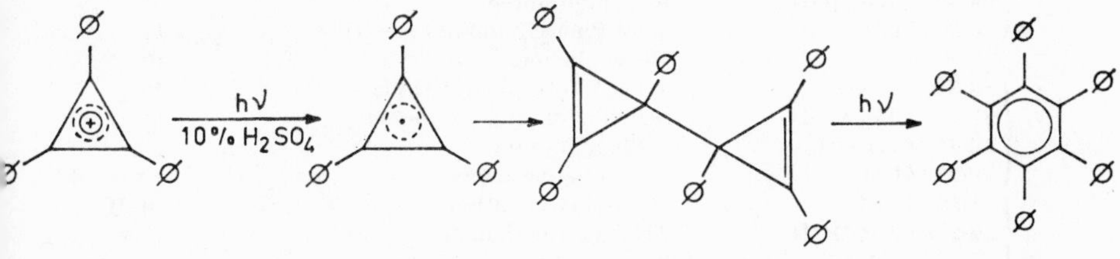

(81)   (83)   (84)   (82)

turn photolyzes to the hexaphenylbenzene. When the radical 83 is generated by other methods, it is found to undergo irreversible dimerization to 84. It is also known that the dimer will undergo photochemical rearrangement to hexaphenylbenzene (Krebs, 1965).

The di-n-propyl cyclopropenyl cation failed to photolyze either in aqueous acid or organic solvents, with or without sensitizers. A possible explanation in the discrepancy between the triphenyl system and this one lies in the calculated energy differences between the cations and their corresponding radicals. In the triphenyl system this energy difference is 0.5β or 16 kcal/mol, while in the di-n-propyl case it is 1.00β or 32 kcal/mol, based on calculated delocalization energies for the two species.

B. *Triphenylmethyl (Trityl) Cation*

The highly stable triphenylmethyl cation (85) is found to undergo a wide variety of photoreactions depending on such variables as the nature of the solvent, the pH of the medium, and the presence or absence of

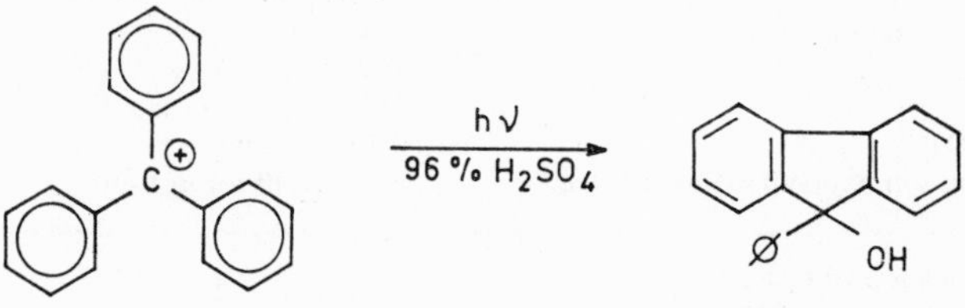

(85)   (86)

TABLE 1

Dependence of Product Formation on Solvent Composition in the Irradiation
of the Trityl Cation (Inert Atmosphere)

Photolysis Medium	Products	Yield (%)
72% H_2SO_4 (aq.)	Dimers, **87**	19·5
	88	22
	and **89**	22
96% H_2SO_4 (aq.)	9-Phenylfluoren-9-ol	60
3·3% H_2SO_4	Triphenylmethane	21
96·5% CH_3COOH	9-Phenylfluorene	22
0·2% H_2O	bis-9-Phenylfluorenyl peroxide	17
	Benzophenone	10
3·3% H_2SO_4	1,1,1,2-Tetraphenylethylene	45
80·1% CH_3COOH	9-Benzyl-9-phenylfluorene	11
16·4% $C_6H_5CH_3$	9-Phenylfluorene	
0·2% H_2O	Triphenylmethane	
1·5% H_2SO_4	Tetraphenylmethane	5–10
96·2% CH_3COOH	Triphenylmethane	
2·3% C_6H_6	9-Phenylfluorene	
	bis-9-Phenylfluorenyl peroxide	
	Benzophenone	
29·4% H_2SO_4	**89**	12
69·4% CH_3COOH	**100**	30
1·2% H_2O		
61% H_2SO_4	**88**	56
36·5% CH_3COOH		
2·5% H_2O		

van Tamelen et al., 1968; van Tamelen et al., 1971.

TABLE 2

Dependence of Product Formation on Solvent Composition in the Irradiation of the
Trityl Cation (Oxygen Present)

Photolysis medium	Products	Yield (%)
3.3% H_2SO_4 (aq.)[a, b]	Benzophenone	37
96.5% CH_3COOH	**101**	30
0.2% H_2O	**102**	11
(carbonium ion concentration 10^{-3} M)		
1% H_2SO_4 (aq.)[a, b]	Benzophenone	
(carbonium ion concentration 10^{-3} M)	**101**	
	102	
	103	
99% H_2SO_4[c]	9-Fluorenone	
(carbonium ion concentration 10^{-5} M)	9-Phenylfluoren-9-ol	

[a] van Tamelen and Cole, 1970.
[b] van Tamelen et al., 1971.
[c] Allen and Owen, 1971.

oxygen (van Tamelen *et al.*, 1968; van Tamelen and Cole, 1971). The results of the reported photolyses are summarized in Tables 1 and 2.

When, for example, ion **85** was photolyzed in 96% sulfuric acid, 9-phenylfluoren-9-ol (**86**) was generated as the single product.

However in 72% H_2SO_4 no **86** was observed, but rather dimers **87**, **88** and **89** were formed. The mechanism for the dimerization reaction is

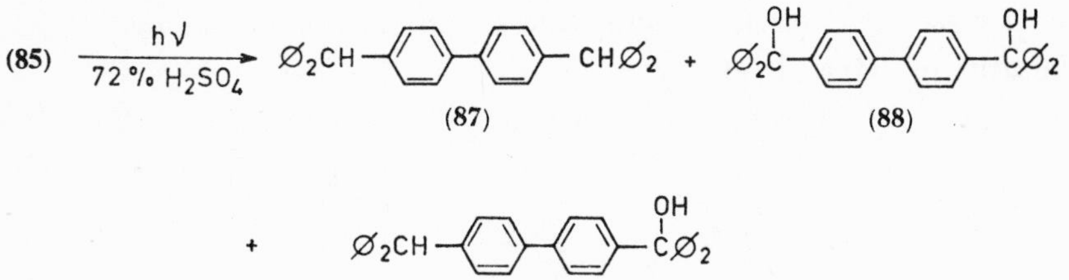

$$(85) \xrightarrow[\text{72\% } H_2SO_4]{h\nu} \varnothing_2CH-\!\!\!\bigcirc\!\!\!-\!\!\!\bigcirc\!\!\!-CH\varnothing_2 + \varnothing_2\overset{OH}{\underset{|}{C}}-\!\!\!\bigcirc\!\!\!-\!\!\!\bigcirc\!\!\!-\overset{OH}{\underset{|}{C}}\varnothing_2$$

(87)   (88)

$$+ \quad \varnothing_2CH-\!\!\!\bigcirc\!\!\!-\!\!\!\bigcirc\!\!\!-\overset{OH}{\underset{|}{C}}\varnothing_2$$

(89)

postulated to proceed via interaction of an excited triplet with the ground state carbonium ion, leading to radical cations **90** and **91**, which then combine to give **92**, the precursor to the observed products.

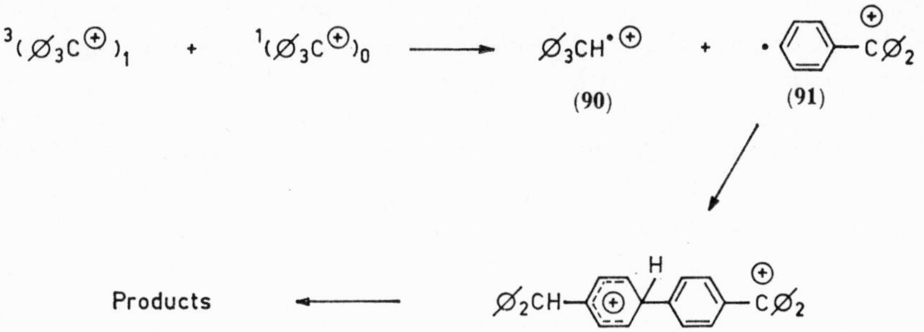

$$^3(\varnothing_3C^{\oplus})_1 \quad + \quad {}^1(\varnothing_3C^{\oplus})_0 \longrightarrow \varnothing_3CH^{\cdot\oplus} + \cdot\!\!\bigcirc\!\!-C\varnothing_2^{\oplus}$$

(90)   (91)

Products $\longleftarrow$ $\varnothing_2CH-\!\!\!\bigcirc\!\!\!\overset{H}{\underset{\oplus}{\bigcirc}}\!\!\!-C\varnothing_2^{\oplus}$

(92)

In weaker acid systems, other reactions involving the triplet state supervene to the exclusion of dimerization. Photolysis of **85** in 3·3% sulfuric acid, 96·5% acetic acid, and 0·2% water gave as products triphenylmethane (**93**), 9-phenylfluorene (**94**), *bis*-9-phenylfluorenyl peroxide (**95**) and benzophenone (**96**). When benzene was present, tetraphenylmethane (**97**) was also formed in addition to the other products. When the triphenylmethyl cation is irradiated in 3·3% H_2SO_4, 80·1% HOAc, 16·4% toluene, and 0·2% H_2O, the products observed were

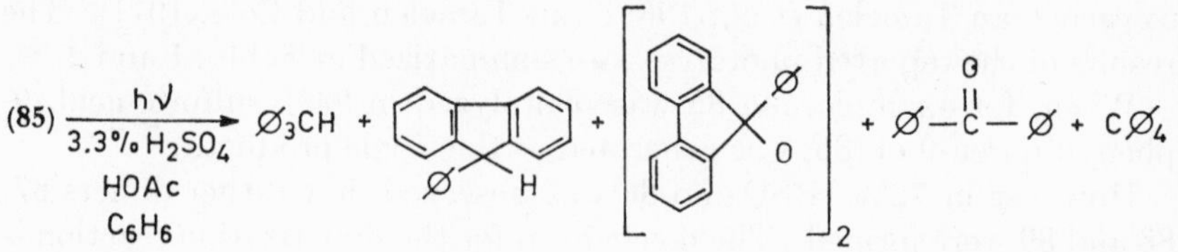

$$(85) \xrightarrow[\substack{3.3\% \ H_2SO_4 \\ HOAc \\ C_6H_6}]{h\nu} \varnothing_3CH + \text{(94)} + \left[\text{(95)} \right]_2 + \varnothing-\overset{\overset{O}{\|}}{C}-\varnothing + C\varnothing_4$$

$$\text{(93)} \qquad \text{(94)} \qquad\qquad \text{(95)} \qquad\qquad \text{(96)} \qquad \text{(97)}$$

1,1,1,2-tetraphenylethane (98), 9-benzyl-9-phenylfluorene (99) and a mixture of (93) and (94).

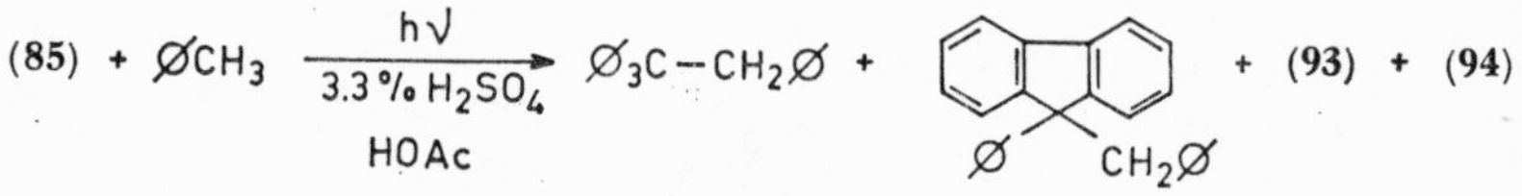

$$(85) + \varnothing CH_3 \xrightarrow[\substack{3.3\% \ H_2SO_4 \\ HOAc}]{h\nu} \varnothing_3C-CH_2\varnothing + \text{(99)} + (93) + (94)$$

$$\text{(98)} \qquad\qquad \text{(99)}$$

It is evident from the nature of the products, especially those formed with toluene present, that the photoreaction in weakly acidic medium involves incursion of a radical species. The complete suppression of reactions leading to the above products, in the presence of oxygen, strongly suggests that it is an excited triplet trityl ion which undergoes reaction. It is postulated that the primary photochemical process is the abstraction of a hydrogen atom by the triplet trityl ion to form the radical cation 90, which was proposed as an intermediate in the dimerization reactions carried out in strong acid (Cole, 1970).

In 29% sulfuric acid in acetic acid (complete ionization) still a different coupling mode is exhibited. The tetraphenylmethane derivative 100 was formed as well as 89. It is worthy of note that in this case coupling

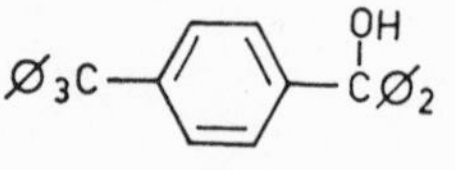

$$\varnothing_3C-\overset{}{\text{⟨⟩}}-\overset{\overset{OH}{|}}{C}\varnothing_2$$

$$\text{(100)}$$

involves the *para* site of the first trityl moiety with the nonannular carbon of the second. At higher mineral acid concentration (61%) in acetic acid no 100 was formed, but rather the *para-para* dimer 88 was observed as the only product. Thus, as the acidity of the acetic acid system is increased, the photoreactions begin to parallel those in aqueous sulfuric acid.

When the trityl cation was photolyzed in 3.3% $H_2SO_4 - 96.5\%$ HOAc -0.2% H_2O in the presence of oxygen, three products were formed:

benzophenone (96), diphenylmethylenedioxybenzene (101) and a substance of empirical formula $C_{21}H_{18}O_4$, which was assigned the structure

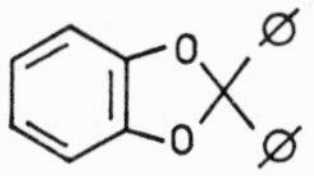

(101) 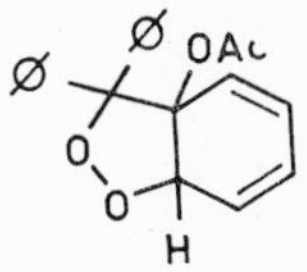   (102)

102 (van Tamelen and Cole, 1970; van Tamelen et al., 1971). At lower acid concentrations (1% H_2SO_4) the aldehyde acetate 103 was observed in addition to the above products.

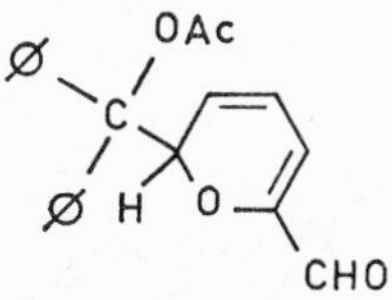

(103)

It is thought unlikely that the mechanism for product formation involves the interaction of triplet carbonium ion with ground state oxygen to give singlet oxygen, followed by reaction of the latter with ground state trityl ions, in view of the electrophilic character of singlet oxygen. In addition, under conditions of complete light absorption the rate of disappearance of 85 was independent of its concentration. The simplest explanation which is consistent with all available data postulates initial capture of oxygen by the triplet trityl ion.

The homolyses of 104 to 105 would be expected on the basis of findings that diaryl peroxides are stable only at very low temperatures, and decompose upon warming to aryloxy radicals (Walling, 1957). The conversion of 105 to 106 finds analogy in the known conversion of triphenylmethoxy radical to the more stable diphenoxymethyl radical (Wieland, 1911). The four-membered peroxide 107 would be expected by precedent to cleave to the ketoaldehyde 108 (Wei et al., 1967; White et al., 1969; Fenical et al., 1969).

In contrast to the findings of van Tamelen et al., whose experiments employed carbonium ion concentrations of $\sim 10^{-3}$ M, irradiation of the trityl ion in somewhat lower concentrations ($\sim 10^{-5}$ M) in 99% sulfuric acid in the absence of hydrogen donors with oxygen present yielded only two products, ions 109 and 110 (Allen and Owen, 1971). An excited triplet carbonium ion is again invoked, which is thought to give rise to the intermediate 111. This species can then lose an allylic hydride ion

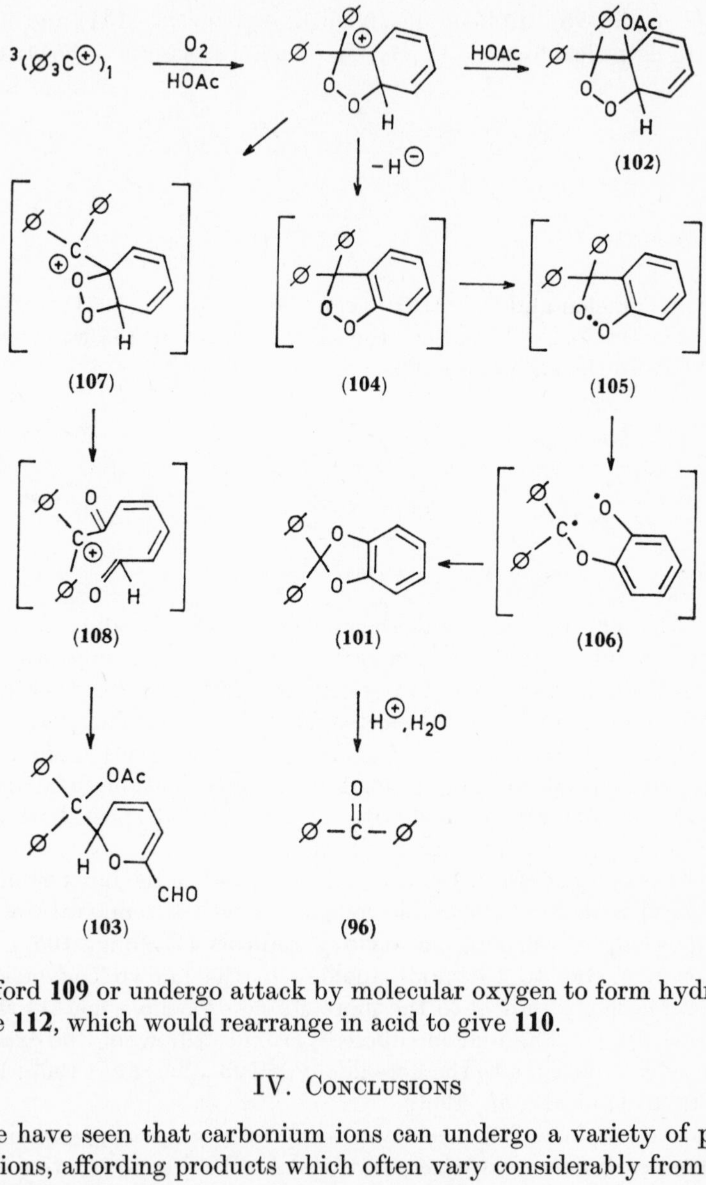

$^3(\varnothing_3C^{\oplus})_1$
O_2
HOAc
HOAc
$-H^{\ominus}$
$H^{\oplus}, H_2O$
OAc
CHO
(102)
(107)
(104)
(105)
(108)
(101)
(106)
(103)
(96)

to afford **109** or undergo attack by molecular oxygen to form hydroperoxide **112**, which would rearrange in acid to give **110**.

IV. Conclusions

We have seen that carbonium ions can undergo a variety of photoreactions, affording products which often vary considerably from those obtained in the photolysis of the corresponding uncharged compounds. The predominant mode of reaction encountered would seem to be isomerization to one or more valence bond isomers, which occurs via a symmetry-allowed disrotatory electrocyclic closure, rather than a $[\sigma 2_a + \pi 2_a]$ cycloaddition in the case of alkylbenzenium ions and pro-

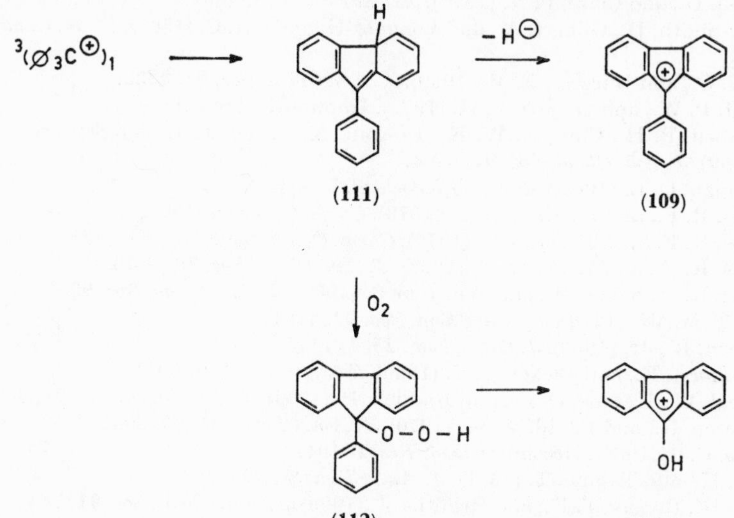

$$^3(\varnothing_3C^{\oplus})_1 \longrightarrow$$

(111)   (109)

$$\downarrow O_2$$

(112)   (110)

tonated cyclohexadienones. Where valence isomerization is precluded, electrocyclic ring opening reactions, or electron transfer and subsequent coupling are observed. In the case of the trityl cation, many pathways may become operative, depending primarily on the acidity of the medium in which the reaction is carried out.

While only the most elementary mechanistic studies have been undertaken, their results indicate that it is the excited singlet which is the reactive species in some instances, while the triplet undergoes reaction in others. To data, there has been little detailed information derived from experiments which would allow one to establish mechanisms or to discuss the lifetimes and energies of the occurring intermediates. Studies have been restricted in large measure to product determination.

However, with the advent of the superacid solvents, a multitude of long-lived stable carbonium ions have been made available for extended study. Clearly, then, the door is open to quantum yield determinations, kinetic treatment, studies on the effects of various solvents and sensitizers and quenching experiments (compare the very recent study by Bethell and Clare, 1972). In short, the photochemistry of carbonium ions is still in its infancy and there is a wealth of information yet to be gained.

REFERENCES

Allen, D. M., and Owen, E. D. (1971). *Chem. Commun.* 848.
Barltrop, J. A., Dawes, K., Day, A. C., and Summers, A. J. H. (1972). *Chem. Commun.* 1240.

Bethell, D., and Clare, P. N. (1972). *J.C.S. Perkin II*, 1464.

Bryce-Smith, D., Gilbert, A., and Longuet-Higgins, H. C. (1967). *Chem. Commun.* 240.

Büchi, G., and Burgess, E. M. (1960). *J. Am. Chem. Soc.* **82**, 4333.

Cabell, P. W., and Hogeveen, H. (1972). Unpublished results.

Campbell, P. H., Chiu, N. W. K., Deugau, K., Miller, I. J., and Sorensen, T. S. (1969). *J. Am. Chem. Soc.* **91**, 6404.

Chapman, O. L. (1963). *Advan. Photochem.* **1**, 353.

Childs, R. F., and Parrington, B. (1970). *Chem. Commun.* 1540.

Childs, R. F., and Taguchi, V. (1970). *Chem. Commun.* 695.

Childs, R. F., and Winstein, S. (1968), *J. Am. Chem. Soc.* **90**, 7146.

Childs, R. F., Sakai, M., and Winstein, S. (1968). *J. Am. Chem. Soc.* **90**, 7144.

Cole, T. M., Jr. (1970). *J. Am. Chem. Soc.* **92**, 4124.

Dauben, H., Jr. (1960). *J. Org. Chem.* **25**, 1442.

Farenhorst, E., and Bickel, A. F. (1966). *Tetrahedron Lett.* 5911.

Fenical, W., Kearns, D. R., and Radlick, P. (1969). *J. Am. Chem. Soc.* **91**, 3396.

Filipescu, N., and Pavlik, J. W. (1970). *J. Am. Chem. Soc.* **92**, 6062.

Foote, C. S. (1968). *Accounts Chem. Res.* **1**, 104.

Hart, H., and Takino, T. (1971). *J. Am. Chem. Soc.* **93**, 720.

Hart, H., Rogers, T. R., and Griffiths, J. (1969). *J. Am. Chem. Soc.* **91**, 754.

Hine, K. E., and Childs, R. F. (1971). *J. Am. Chem. Soc.* **93**, 2323.

Hine, K. E., and Childs, R. F. (1972). *Chem. Commun.* 145.

Hogeveen, H., and Gaasbeek, C. J. (1970). *Rec. Trav. Chim.* **89**, 1079.

Hogeveen, H., and Kwant, P. W. (1972). *Tetrahedron Lett.* 3197.

Hogeveen, H., Kellogg, R. M., and Kuindersma, K. A. (1972). Unpublished results.

Hurst, J. J., and Whitham, G. H. (1963). *J. Chem. Soc.* 710.

Joussot-Dubien, J., and Houdard-Pereyre, J. (1969). *Bull. Soc. Chim. Fr.* 2619.

Jutz, C., and Voithenleiter, F. (1964). *Chem. Ber.* **97**, 29.

Kaplan, L., Pavlik, J. W., and Wilzbach, K. E. (1972). *J. Am. Chem. Soc.* **94**, 3283.

Knapczyk, J. W., Lubinowski, J. J., and McEwen, W. E. (1972). *Tetrahedron Lett.* 3739.

Koptyug, V. A., Kuzubova, L. I., Isaev, I. S., and Mamatyuk, V. I. (1969). *Chem. Commun.* 389.

Krebs, A. W. (1965). *Angew. Chem. Int. Ed. Engl.* **4**, 10.

Lustgarten, R. K., Brookhart, M., and Winstein, S. (1967). *J. Am. Chem. Soc.* **89**, 6350.

Meinwald, J., and Mazzocchi, P. H. (1966). *J. Am. Chem. Soc.* **88**, 2850.

Mukai, T., Tezuka, T., and Akasaki, Y. (1966). *J. Am. Chem. Soc.* **88**, 5025.

Noyori, R., Ohnishi, Y., and Kato, M. (1972). *J. Am. Chem. Soc.* **94**, 5105.

Parrington, B., and Childs, R. F. (1970). *Chem. Commun.* 1581.

Shuster, D. I., and Sussman, D. H. (1970). *Tetrahedron Lett.* 1657.

Shuster, D. I., Nash, M. J., and Kantor, M. L. (1964). *Tetrahedron Lett.* 1375.

Sorenson, T. S. (1965). *J. Am. Chem. Soc.* **87**, 5075.

Takino, T., and Hart, H. (1970). *Chem. Commun.* 450.

van Tamelen, E. E., and Cole, T. M., Jr. (1970). *J. Am. Chem. Soc.* **92**, 4123.

van Tamelen, E. E., and Cole, T. M., Jr. (1971). *J. Am. Chem. Soc.* **93**, 6158.

van Tamelen, E. E., Cole, T. M., Jr., Greeley, R., and Schumacher, H. (1968). *J. Am. Chem. Soc.* **90**, 1372.

van Tamelen, E. E., Greeley, R. H., and Schumacher, H. (1971). *J. Am. Chem. Soc.* **93**, 6151.

Walling, C. (1957). "Free Radicals in Solution", Wiley, New York, p. 466.

Wei, K., Mani, J., and Pitts, J. N., Jr. (1967). *J. Am. Chem. Soc.* **89**, 4225.

White, E. H., Wieko, J., and Rosewell, D. F. (1969). *J. Am. Chem. Soc.* **91**, 5194.

Wieland, H. (1911). *Ber.* **44**, 2533.

Wilzbach, K. E., and Rausch, D. J. (1970). *J. Am. Chem. Soc.* **92**, 2178.

Woodward, R. B., and Hoffmann, R. (1970). "The Conservation of Orbital Symmetry", Academic Press, New York.

Yoshida, L., Sugimoto, T., and Yenida, S. (1971). *Tetrahedron Lett.* 4259.

PHYSICAL PARAMETERS FOR THE CONTROL OF ORGANIC ELECTRODE PROCESSES

M. FLEISCHMANN AND D. PLETCHER

Department of Chemistry, The University, Southampton SO9 5NH, England

I. Introduction	155
II. Electrode Potential	156
A. The Electron Transfer Process	157
B. The Adsorption Equilibria	165
C. The Electrode Surface	171
III. The Solution Environment	172
A. Basic Requirements	173
B. The Environment as a Reactant	174
C. pH Effects	178
D. Double Layer and Adsorption Effects	184
IV. Electrode Material	191
V. Substrate Concentration	198
VI. Temperature	201
VII. Pressure	204
VIII. Structural Effects	206
A. The Rates of Simple Electrode Reactions	206
B. Reactions of Intermediates	210
C. Reaction Coordinates	211
IX. Cell Design	213
References	220

I. INTRODUCTION

IN recent years several factors have combined to stimulate renewed interest in the application of electrochemical methods to organic synthesis. Amongst these one might list the development of electronic instrumentation for the control and investigation of electrode reactions which has led to an improved understanding of these processes; an increased interest in the modes of reaction of energetic intermediates such as the radical ions and ions which may be generated at the electrode surface; the possibility, at least in principle, of designing novel, selective reactions which, moreover would involve the electron rather than complex chemical reagents, thereby leading to simpler and cleaner reaction systems. Although a large number of interesting reactions have been reported (see reviews and books: Adams, 1969a; Anderson *et al.*, 1969; Baizer, 1969; Baizer and Petrovich, 1970; Beck, 1970; Bewick and Pletcher, 1971, 1972; Brown and Harrison, 1969; Cauquis, 1966; Chang *et al.*, 1971; Eberson, 1968; Eberson and Schäfer, 1971; Fichter, 1942;

Fleischmann and Pletcher, 1969, 1971; Lund, 1967; Mann and Barnes, 1970; Peover, 1967, 1971; Wawzonek, 1967; Weinberg and Weinberg, 1968) and in spite of the fact that reactive intermediates may readily be produced from inert starting materials, electrochemical methods have still not found general application.

One possible reason for the reluctance of non-electrochemists to venture into this field is that in contrast to the electrochemists' claim that controlled potential electrolysis offers a method for the selective introduction of energy into molecules, many electrode reactions carried out at a controlled potential have still been reported to give low yields and a diversity of products. The electrode potential is, however, only one of several variables and the lack of selectivity in the electrode process may be attributed to a failure to understand and to control all the parameters of the overall electrode reaction.

When considering the effect of each parameter, it must be remembered that an electrode process consists of a sequence of steps. For example in the simplest possible electrode reaction it is necessary for the reactant to diffuse from the bulk of the solution to the electrode surface, for the electron transfer to occur and for the product to return to the bulk; in more complex electrode reactions there may be a chemical step preceding or following the electron transfer and it may be necessary to consider the adsorption of the reactant or the product on the electrode surface. Furthermore the product of the electrode reaction will be, in general, a reactive intermediate which will undergo further reaction steps. Each parameter may affect one or more of these stages of the overall reaction path and the particular choice of conditions will control the nature of the rate-controlling step and thereby the steady state current. A successful synthesis will demand careful choice of the conditions to ensure not only selective generation of the intermediates in the electrode reaction but also selective reactions of these species.

This chapter is intended to outline the present state of our knowledge about the factors which affect the mechanism, rate and products of an electrode reaction and to show, in turn, how the various parameters affect the individual steps in the overall reaction. In addition it will be shown how many of these parameters, once fully understood, may be turned to an advantage and may be used to design further novel electrosynthetic processes.

II. Electrode Potential

The potential of the working electrode is important to the overall electrode process largely because of its effect on the electron transfer step. As will be discussed later, the electrode potential may also affect the reaction by changing the adsorption characteristics of the starting material, intermediates, or the product of the electrode reaction and, particularly in the case of solid electrodes, by controlling the state of the electrode surface. In most electrode reactions, however, it is its effect on the electron transfer step which is the predominant reason for the need to control the electrode potential.

A. The Electron Transfer Process

Firstly, it is the electrode potential which determines whether sufficient energy is being supplied for the electron transfer to occur. If we consider the simple electrode reaction (omitting charge designation for the oxidized and reduced species O and R)

$$O + e \rightleftharpoons R \tag{1}$$

coupled to a further reaction, normally that at a hydrogen electrode,

$$H^+ + e \rightleftharpoons \tfrac{1}{2}H_2 \tag{2}$$

so that the overall cell reaction is

$$O + \tfrac{1}{2}H_2 \rightleftharpoons R + H^+ \tag{3}$$

then the standard free energy of the reaction, ΔG^0, is related to the reversible potential, E^0, by the equation

$$\Delta G^0 = -FE^0 \tag{4}$$

where F is the Faraday. If n electrons are involved in reactions (1) and (2) then

$$\Delta G^0 = -nFE^0 \tag{5}$$

In these reactions, (2) is the process taking place at the reference electrode which therefore determines the potential scale. In practice other reference electrodes, such as the saturated calomel electrode are frequently used but the data are normally expressed on the hydrogen scale.

The equilibrium (1) at the electrode surface will lie to the right, i.e. the reduction of O will occur if the electrode potential is set at a value more cathodic than E^0. Conversely, the oxidation of R would require the potential to be more anodic than E^0. Since the potential range in certain solvents can extend from $-3 \cdot 0$ V to $+3 \cdot 5$ V, the driving force for an oxidation or a reduction is of the order of 3 eV or 260 kJ mol^{-1} and experience shows that this is sufficient for the oxidation and reduction of most organic compounds, including many which are resistant to chemical redox reagents. For example, the electrochemical oxidation of alkanes and alkenes to carbonium ions is possible in several systems

(Bertram *et al.*, 1971; Clark *et al.*, 1972; Coleman *et al.*, 1972; Fleischmann and Pletcher, 1968).

It is instructive to draw up a free-energy cycle for the cell reaction (3) so as to illustrate the dominant energy terms in the single electrode reaction (1):

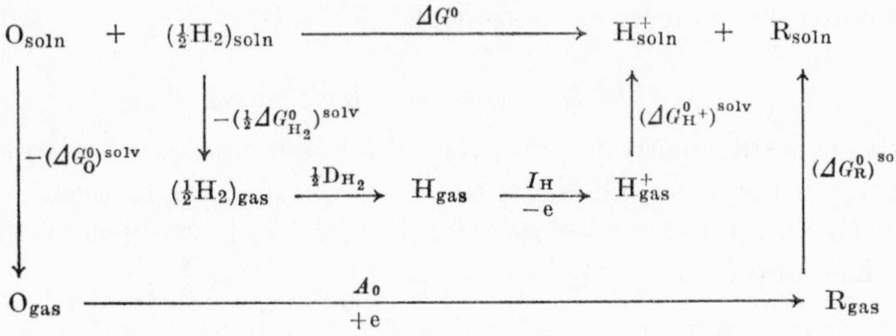

$$O_{soln} + (\tfrac{1}{2}H_2)_{soln} \xrightarrow{\;\;\Delta G^0\;\;} H^+_{soln} + R_{soln}$$

where A_0 is the electron affinity of O, D_{H_2} the dissociation energy of H_2 and I_H the ionization potential of H while the remaining terms are solvation free energies. It follows that

$$\Delta G^0 = A_0 + (\Delta G^0_R)^{solv} - (\Delta G^0_O)^{solv} + [\tfrac{1}{2}D_{H_2} + I_H + (\Delta G^0_{H^+})^{solv} - (\Delta G^0_{H_2})^{solv}] \tag{6}$$

and this equation will determine the cell potential (see equation (4)). It is interesting to note that for a series of compounds, such as polynuclear hydrocarbons, the potentials of the radical anion/hydrocarbon couples correlate linearly with their electron affinities or comparable quantities such as the energy of the lowest unfilled molecular orbital found from quantum mechanical calculations (Peover, 1967, 1971); likewise oxidation potentials correlate with the first ionization potential (Peover, 1967). This suggests that for a series of compounds $(\Delta G^0_O)^{solv}$ and $(\Delta G^0_R)^{solv}$ are constant or vary in a linear manner with the electron affinity (or ionization potential): oxidation and reduction potentials are therefore simple forms of a linear free energy relation.

As in chemical systems, however, the requirement that the reaction is thermodynamically favourable is not sufficient to ensure that it occurs at an appreciable rate. In consequence, since the electrode reactions of most organic compounds are irreversible, i.e. slow at the reversible potential, it is necessary to supply an overpotential, $\eta = E - E^0$, in order to make the reaction proceed at a conveniently high rate. Thus, secondly, the potential of the working electrode determines the kinetics of the electron transfer process.

When the potential of the electrode is E with respect to a reference electrode, the potential of the surface of the working electrode will be ϕ_m

with respect to the solution. Making the simplest assumption that the potential ϕ_m is established between the plane of closest approach of the reactant and the electrode surface, a fraction $\alpha n F \phi_m$ of the maximum work $n F \phi_m$ will affect the energy of activation of the forward reaction of (7),

$$\text{O} + n e \underset{k_2}{\overset{k_1}{\rightleftarrows}} \text{R} \qquad (7)$$

while the fraction $(1 - \alpha) n F \phi_m$ will affect the energy of activation of the reverse process. Then the net rate is given by

$$\text{Rate} = A k_1 C_0 \exp\left(\frac{-\alpha n F}{RT} \phi_m\right) - A k_2 C_R \exp\left(\frac{(1-\alpha) n F}{RT} \phi_m\right) \qquad (8)$$

or, in terms of the current,

$$i = n F A k_1 C_0 \exp\left(\frac{-\alpha n F}{RT} \phi_m\right) - n F A k_2 C_R \exp\left(\frac{(1-\alpha) n F}{RT} \phi_m\right) \qquad (9)$$

where A is the area of the electrode and C_0 and C_R are the concentrations of O and R respectively. α is known as the transfer coefficient.

Equation (9) shows that for a reduction, when ϕ_m is sufficiently negative,

$$\ln i_\text{C} = \ln n F A k_1 C_0 - \frac{\alpha n F}{RT} \phi_m \qquad (10)$$

while for an oxidation, when ϕ_m is sufficiently positive,

$$\ln i_\text{A} = \ln n F A k_2 C_R + \frac{(1-\alpha) n F}{RT} \phi_m \qquad (11)$$

and in both cases the current increases logarithmically with the applied potential.

Equation (9) is often reduced to the equivalent form

$$i = i_0 A \left[\exp\left(\frac{-\alpha n F}{RT} \eta\right) - \exp\left(\frac{(1-\alpha) n F}{RT} \eta\right) \right] \qquad (12)$$

where i_0, the exchange current, is the partial anodic and cathodic currents at the reversible potential (the net current is zero at this potential). It is given by

$$i_0 = k^0 C_0^{(1-\alpha)} C_R^\alpha \qquad (13)$$

where k^0 is the rate constant at the standard electrode potential (see also Damaskin, 1967; Koryta et al., 1966; Vetter, 1967).

The potential ϕ_m of the electrode with respect to the solution can be expressed in terms of the reversible potential ϕ_m^r and the overpotential, η

$$\phi_m = \phi_m^r + \eta \qquad (14)$$

Substitution in (9) gives

$$i = nFAk_1 C_0 \exp\left(\frac{-\alpha nF}{RT}\phi_m^r\right)\exp\left(\frac{-\alpha nF}{RT}\eta\right)$$
$$- nFAk_2 C_R \exp\left(\frac{(1-\alpha)\,nF}{RT}\phi_m^r\right)\exp\left(\frac{(1-\alpha)\,nF}{RT}\eta\right) \qquad (15)$$

At the reversible potential $\eta = 0$, $i = 0$, and

$$k_1 C_0 \exp\left(\frac{-\alpha nF}{RT}\phi_m^r\right) = k_2 C_R \exp\frac{(1-\alpha)\,nF}{RT}\phi_m^r = k = \frac{i_0}{nF} \qquad (16)$$

Therefore equation (15) may be rewritten

$$i = nFkA\left[\exp\left(\frac{-\alpha nF}{RT}\eta\right) - \exp\left(\frac{(1-\alpha)\,nF}{RT}\eta\right)\right] \qquad (17)$$

$$= i_0 A\left[\exp\left(\frac{-\alpha nF}{RT}\eta\right) - \exp\left(\frac{(1-\alpha)\,nF}{RT}\eta\right)\right] \qquad (18)$$

Since ϕ_m^r can be expressed in terms of the Nernst equation (neglecting activity coefficients)

$$\phi_m^r = \phi_m^{r,0} + \frac{RT}{nF}\ln\frac{C_0}{C_R} \qquad (19)$$

where $\phi_m^{r,0}$ is the standard electrode potential (i.e. $C_0 = C_R$), by substituting (19) in (16)

$$k = \frac{i_0}{nF} = k^0 C_0^{(1-\alpha)} C_R^{\alpha} \qquad (20)$$

where

$$k^0 = k_1 \exp\left(\frac{-\alpha nF}{RT}\phi_m^{r,0}\right) = k_2 \exp\left(\frac{(1-\alpha)\,nF}{RT}\phi_m^{r,0}\right) \qquad (21)$$

It is important to note for a later section that a simple dependence of the cathodic current (or anodic current) on the concentration $C_0(C_R)$ is obtained if the experiment is carried out at constant electrode potential [see equations (10) and (11)] but not if the experiment is carried out at constant overpotential since the value of the reversible potential depends on the ratio of C_0/C_R in solution [see equation (19)]. The significance of k^0 is, however, exact.

The correct potential for a preparative electrolysis is normally chosen by inspection of a steady state current–potential (i–E) curve. Figure 1 shows a typical i–E curve for the reduction of anthracene at a mercury cathode in dimethylformamide (Peover *et al.*, 1963); the curve shows two reduction waves. In the potential range where the current rises with variation of the potential, the rate of an electron transfer process is increasing while in the plateau regions the rate of the electron transfer has reached a value where another reaction, either a preceding chemical reaction or, more commonly, the diffusion of the electroactive species to the electrode surface, has become the rate-determining step. The distinction can readily be made by use of other electrochemical techniques, e.g., rotating disc electrode (Adams, 1969a; Piekarski, 1971; Riddiford, 1968), cyclic voltammetry (Adams, 1969a; Brown and Large, 1971), potential step (Delahay, 1961). Furthermore, if the plateau currents are controlled by diffusion, the ratio of the heights of the reduction waves in the $i–E$ curve is a reflection of the ratio of the number of electrons, n, transferred in the processes. [An absolute determination of n requires the use of controlled potential coulometry (Meites, 1971) and an analysis of the slope of the

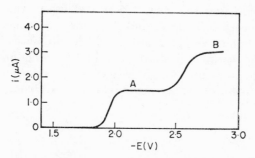

FIG. 1. $i–E$ curve for the reduction of anthracene at a dropping mercury electrode in acetonitrile containing 0·1 M tetraethylammonium perchlorate.

rising portion of the $i–E$ curve makes it possible to distinguish whether the process is reversible or irreversible (Meites, 1965). In the example cited in Fig. 1, it is clear that two electron transfer processes occur within the potential range of the solvent and it could be shown that, in fact, they are the addition of one electron to the neutral substrate followed by the addition of a further electron to the radical anion produced. The addition of the first electron is reversible while the second reaction is irreversible. Since different products may be expected from the addition of one or two electrons to anthracene and it is clearly desirable to form the products at the maximum rate, i.e., at the highest current, the preparative electrolyses would normally be carried out at a potential around A or B, depending on which product is desired.

The oxidation or reduction of many organic substrates may lead to the formation of two or more reactive intermediates and the products which are isolated depend on the reactions of these intermediates in the environment of the electrode. It is primarily the electrode potential which determines which intermediate is formed and also the rate at which it is obtained. Thus it is clearly of prime importance to control the electrode potential if a highly selective process is desired. This concept is not new and dates from the work of Haber (1898). He studied the reduction of nitrobenzene in aqueous acid and showed that the product was phenylhydroxylamine at low negative potentials and aniline at more negative potentials.

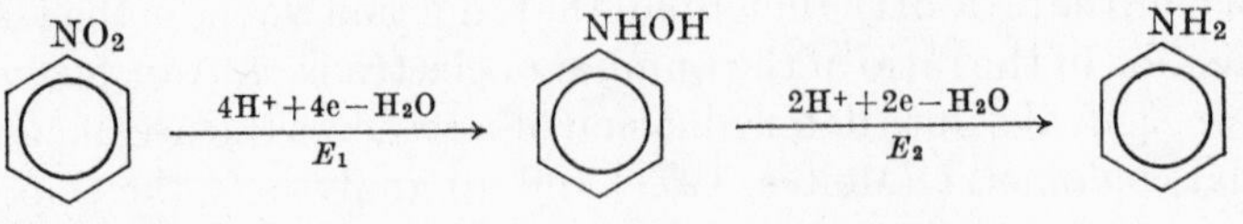

The development and the very widespread use of the polarographic technique to record i–E curves and the more recent designing of electronic devices known as potentiostats which automatically control the potential of the working electrode at a pre-set value has led to many examples in the literature of organic electrode reactions whose products depend on the potential. Some examples are cited below:

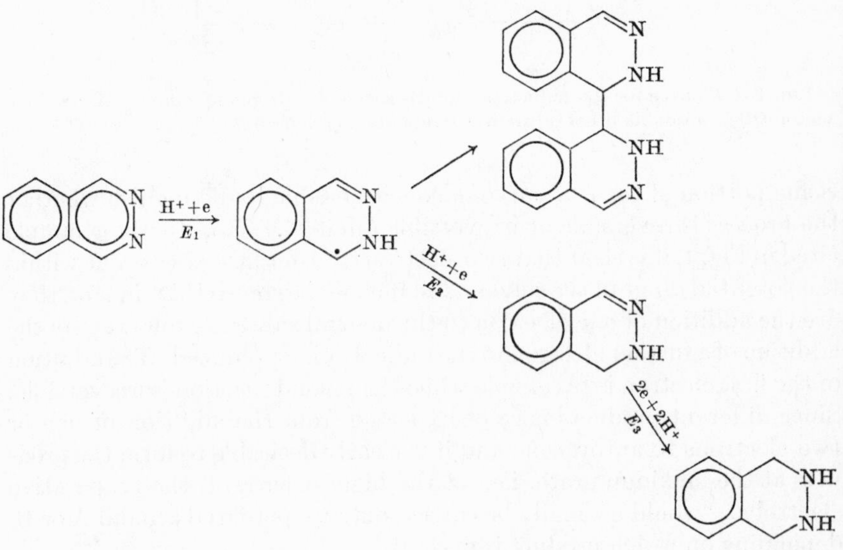

(Lund, 1966)

$$CCl_3.CO.OH \xrightarrow[E_1]{2e + H^+ - Cl^-} CHCl_2.CO.OH \xrightarrow[E_2]{2e + H^+ - Cl^-}$$

$$CH_2Cl.CO.OH \xrightarrow[E_3]{2e + H^+ - Cl^-} CH_3.CO.OH$$

(Elving and Hilton, 1952)

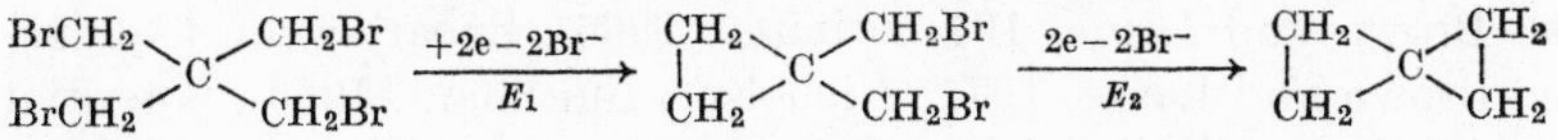

$$\begin{array}{ccc} BrCH_2 \diagdown & & \diagup CH_2Br \\ & C & \\ BrCH_2 \diagup & & \diagdown CH_2Br \end{array} \xrightarrow[E_1]{+2e-2Br^-} \begin{array}{ccc} CH_2 \diagdown & & \diagup CH_2Br \\ & C & \\ CH_2 \diagup & & \diagdown CH_2Br \end{array} \xrightarrow[E_2]{2e-2Br^-} \begin{array}{ccc} CH_2 \diagdown & & \diagup CH_2 \\ & C & \\ CH_2 \diagup & & \diagdown CH_2 \end{array}$$

(Rifi, 1971)

$$\begin{array}{ccc} C_6H_5 \diagdown & & \diagup C_6H_5 \\ & C=C & \\ C_6H_5 \diagup & & \diagdown Br \end{array} \xrightarrow[E_1]{2e+H^+-Br^-} \begin{array}{ccc} C_6H_5 \diagdown & & \diagup C_6H_5 \\ & C=C & \\ C_6H_5 \diagup & & \diagdown H \end{array} \xrightarrow[E_2]{2e+2H^+}$$

$$\begin{array}{c} C_6H_5 \diagdown \\ \quad\quad CH-CH_2-C_6H_5 \\ C_6H_5 \diagup \end{array}$$

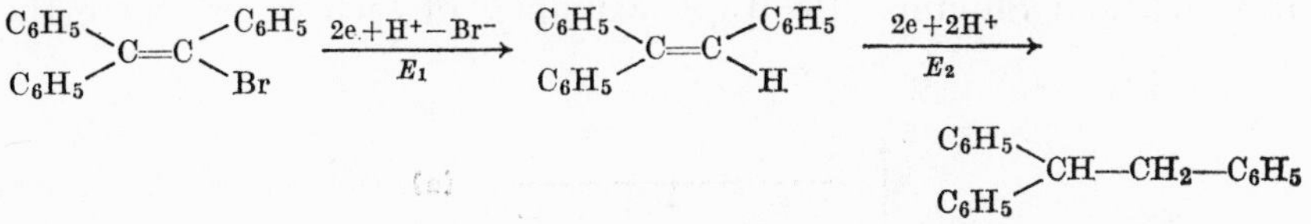

(Miller and Riekena, 1969)

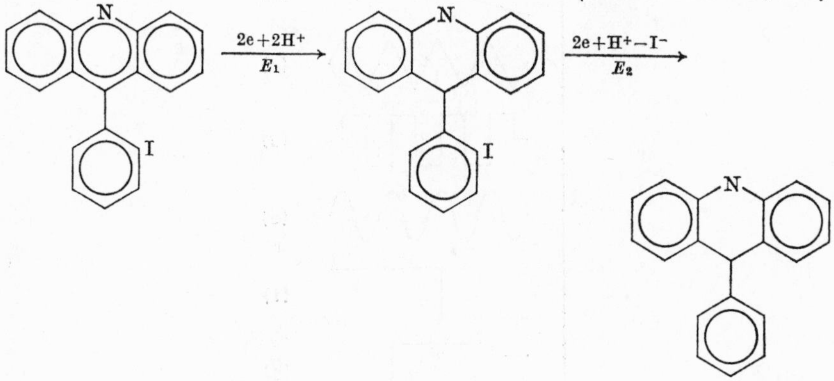

$$\xrightarrow[E_1]{2e+2H^+}$$

$$\xrightarrow[E_2]{2e+H^+-I^-}$$

(Fry et al., 1970)

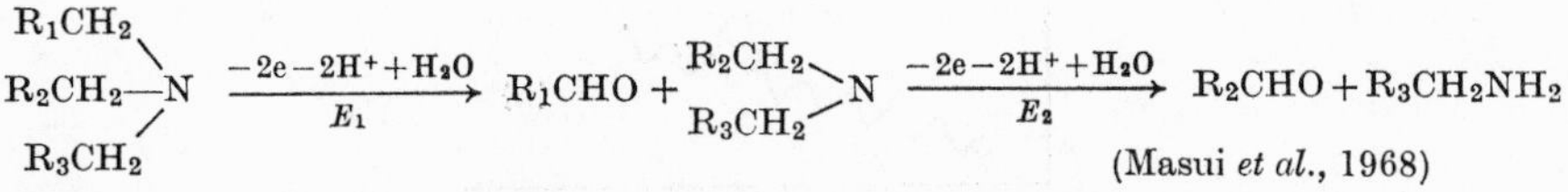

$$\begin{array}{c} R_1CH_2 \diagdown \\ R_2CH_2-N \\ R_3CH_2 \diagup \end{array} \xrightarrow[E_1]{-2e-2H^++H_2O} R_1CHO + \begin{array}{c} R_2CH_2 \diagdown \\ \quad N \\ R_3CH_2 \diagup \end{array} \xrightarrow[E_2]{-2e-2H^++H_2O} R_2CHO + R_3CH_2NH_2$$

(Masui et al., 1968)

$$\begin{array}{c} C_6H_5 \diagdown \\ \quad N-NH_2 \\ C_6H_5 \diagup \end{array} \xrightarrow[E_1]{-2e-H^+} \begin{array}{c} C_6H_5 \diagdown \overset{+}{} \\ \quad N=NH \\ C_6H_5 \diagup \end{array} \xrightarrow{C_6H_5CH=CH_2 \atop H^+}$$

$$\xrightarrow[E_1]{-2e-2H^+}$$

$$\xrightarrow[E_2]{-2e-H^+}$$

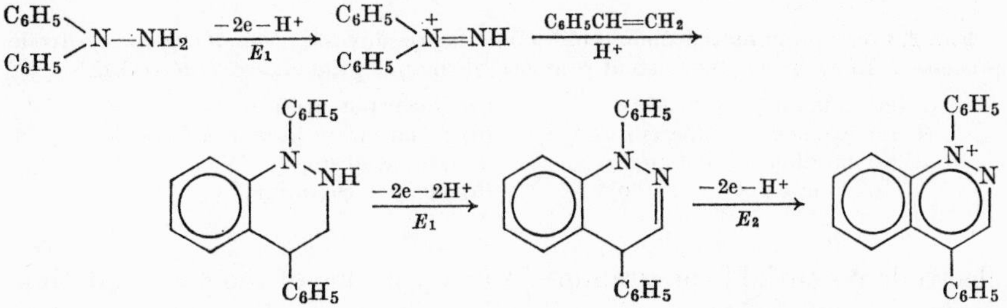

(Cauquis and Genies, 1970)

In the majority of examples in the literature, the desired product is obtained by applying a single fixed potential. It is also possible, however, to programme the electrode potential to change in a pre-determined manner, and this is commonly done in studies of reaction intermediates and the kinetics of electrode processes. Thus, cyclic voltammetry (Adams,

1969a; Brown and Large, 1971; Meites, 1965), linear sweep (Adams, 1969a; Brown and Large, 1971; Nicholson and Shain, 1966), potential step (Gerischer and Vielstich, 1955), double potential step (Christie, 1967; Schwarz and Shain, 1965) and faradaic rectification (Sluyters-Rehbach and Sluyters, 1970) are examples of techniques where the

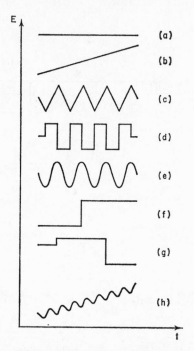

FIG. 2. Common potential–time profiles used for the investigation of organic electrode processes. In each case the current response to the potential change is recorded.

(a) constant potential   (b) linear potential sweep
(c) cyclic voltammetry   (d) square wave pulse electrolysis
(e) faradaic rectification   (f) potential step
(g) double potential step   (h) a.c. polarography.

electrode potential is programmed to vary as one of the potential–time profiles shown in Fig. 2. It is less common for such a potential–time profile to be used to control a preparative electrolysis although such an approach would greatly extend the scope of organic electrosynthesis. Mechanistic data in the literature show that many novel one step processes could be constructed by using a square wave potential–time profile. An example which has been reported is the synthesis of benzidine from nitrobenzene (Fleischmann et al., 1964): at one potential the nitrobenzene is reduced to phenylhydroxylamine while at the more anodic potential some of this intermediate is reoxidized to nitrosobenzene. The phenylhydroxylamine and the nitrosobenzene can then interact to form a product which is further reducible.

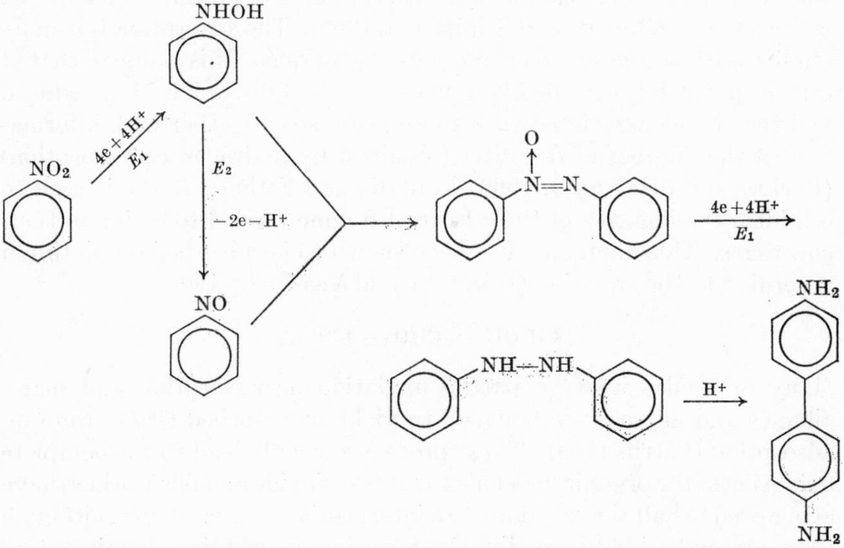

In such a synthesis the lengths of the pulses are variable as well as the potentials of the square wave. Recently a potential–time profile has been used to maintain the activity of an electrode during the oxidation of organic compounds (Clark *et al.*, 1972): at a steady potential the current for the oxidation process was observed to fall, but a periodic short pulse to cathodic potentials was sufficient to prevent this decrease in electrode activity.

B. *The Adsorption Equilibria*

Many organic electrode processes require the adsorption of the electroactive species at the electrode surface before the electron transfer can occur. This adsorption may take the form of physical or reversible chemical adsorption, as has been commonly observed at a mercury/water interface, or it may take the form of irreversible, dissociative chemical adsorption where bond fracture occurs during the adsorption process and often leads to the complete destruction of the molecule. This latter type of adsorption is particularly prevalent at metals in the platinum group and accounts for their activity as heterogeneous catalysts and as fuel cell electrodes (Bockris and Srinivasan, 1968; Petrii, 1969). In all types of adsorption, the coverage of the electrode by the organic species is strongly dependent on the potential of the metal surface.

The dissociative chemisorption of many organic molecules at solid electrodes has been studied in aqueous solution and at various temperatures and pressures in connection with the effort to produce efficient fuel cell electrodes (Bockris and Srinivasan, 1969). The adsorption is usually studied with potential sweep and pulse techniques. It is assumed that at certain potentials the adsorbed phase may be fully oxidized or reduced and the charge associated with these processes, together with information of the number of free sites (obtained by hydrogen co-adsorption) (Bockris and Srinivasan, 1969; Franklin and Sothern 1954), is used to establish the identity of the adsorbed fragments and to estimate their coverages. Thus methanol has been shown to be adsorbed on platinum according to the equation (Bagotzky and Vassilyev, 1967)

$$CH_3OH \rightarrow CHO_{ADS} + 3H_{ADS}$$

Many molecules undergo partial oxidation on adsorption and many alkanes and alkenes are believed to yield an adsorbed CHO group on adsorption (Petrii, 1968). These processes usually lead to the complete oxidation of the organic molecule to carbon dioxide and few workers have attempted to halt the reaction at an intermediate stage. Hence, although there are undoubtedly possibilities for using dissociative chemisorption for synthetic reactions, this chapter will not consider these processes further.

Non-dissociative adsorption has been studied for a wide range of organic molecules at a mercury surface under conditions where electron transfer does not occur (Delahay, 1966). The coverage, θ, of the surface by the organic species is usually obtained by measuring the change in the interfacial tension–potential curves with the activity in the solution of the species being adsorbed or in a non-thermodynamic manner from capacitance–potential plots. The θ–E data may be interpreted in terms of an isotherm; the simplest isotherm is obtained when the behaviour of the adsorbate is "ideal", i.e., there is no interaction between neighbouring molecules, circumstances which arise only for low coverages. The ideal isotherm is

$$\theta = \beta a \tag{22}$$

where a is the activity of the organic molecule in solution and β is given by equation (23).

$$\beta = \exp\left[-\frac{\Delta G^0_{ADS}}{RT}\right] \tag{23}$$

The free energy of adsorption, ΔG^0_{ADS} is potential-dependent and, in the case of an adsorbate which is an uncharged molecule,

$$\Delta G^0_{ADS} = X + Y(E - E_{MAX})^2 \qquad (24)$$

where X and Y are constants and E_{MAX} is the potential where the coverage is a maximum, normally the potential at which the surface is uncharged. This quadratic dependence corresponds to a simple dielectric model of the interface where part of the layer of solvent adjacent to the electrode is displaced by a material with a different permittivity. With increasing

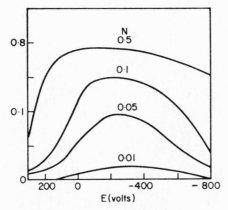

FIG. 3. Electrocapillary curves for various concentrations of butanol in water. (Taken from Bockris *et al.*, 1963.)

magnitude of $(E - E_{MAX})$ the adsorbate, which is normally less polar than the solvent, is progressively expelled. Figure 3 shows a θ–E curve obtained from electrocapillary measurements on an aqueous solution containing butanol and it illustrates the typical quadratic form of these curves (Bockris *et al.*, 1963). This ideal adsorption isotherm is seldom found in practice since there is normally some lateral interaction between neighbouring adsorbed molecules and it is usually necessary to use a more complex isotherm. Thus the Virial and Frumkin isotherms reflect differences in the type and extent of this lateral interaction (Delahay, 1966).

The capacitance is a readily measured interfacial property and it gives qualitative information on the adsorption of species at the electrode surface. Since the surface charge density, q, is a function of the potential and of coverage, the measured capacitance may be expressed as the sum of a "true" (high frequency) capacitance and an adsorption pseudo-capacitance, i.e. $q = f(E, \theta)$ and hence

$$dq = \left(\frac{\partial q}{\partial E}\right)_\theta \partial E + \left(\frac{\partial q}{\partial \theta}\right)_E \partial \theta \qquad (25)$$

$$C = \frac{\partial q}{\partial E} = \left(\frac{\partial q}{\partial E}\right)_\theta + \left(\frac{\partial q}{\partial \theta}\right)_E \frac{\partial \theta}{\partial E} = C_{h,f} + C_\theta \qquad (26)$$

The adsorption pseudo-capacitance, C_θ, is dominated by the factor $\partial\theta/\partial E$ and hence a plot of C_θ versus E gives direct information about the coverage. Figure 4 shows a C_θ–E plot for aniline at mercury in aqueous solu-

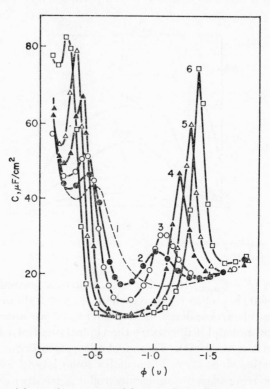

Fig. 4. Differential capacitance–potential curves for various concentrations of aniline in 1·0 M aqueous potassium chloride and at a mercury electrode-frequency 400 Hz.

Curve	1	2	3	4	5	6
Conc.	0	10^{-2} M	2×10^{-2} M	5×10^{-2} M	10^{-1} M	2×10^{-1} M

(Taken from Frumkin, 1966.)

tion (Frumkin, 1966). The peaks in this curve have been interpreted in terms of desorption and reorientation of the adsorbed molecules.

The techniques described above are not entirely suitable for the study of adsorption when a faradaic current is flowing. Hence, little fully substantiated information is available concerning the adsorption of the electroactive species under conditions where a synthetic reaction is occurring. Moreover, once electron transfer occurs, the situation is complicated by the fact that intermediates and products may compete for sites on the electrode surface. The above picture, however, remains useful and indicates the way in which the coverage of the surface by the starting material will vary with potential. Furthermore, coverages by the reactant measured at potentials just prior to the onset of electron transfer have been extrapolated into the potential region where the synthetic reaction occurs and used to elucidate the kinetics of the process (Conway et al., 1968).

More usually, the participation of adsorbed reactants and intermediates is inferred indirectly from kinetic data (Bockris, 1954). Thus the observation of reactions having a low or zero order with respect to the reactant concentration implies adsorption of the reactant. A reaction scheme

$$O + e \rightleftharpoons R_{ADS}$$

$$R_{ADS} + R_{ADS} \xrightarrow[\text{rate determining}]{k} R_2$$

would give an intermediate whose surface concentration θ would be given at low concentrations by $\partial \log_{10} \theta / \partial E = (60 \text{ mV})^{-1}$. Therefore further reaction of the intermediates with a rate $k\theta^2$ will give $\partial \log_{10} i / \partial E = (30 \text{ mV})^{-1}$ and a reaction order of 2; for a slow one-electron transfer step such a "Tafel slope" could only be explained by making the invalid assumption $\alpha > 1$ in equation (9). In practice such simple behaviour is seldom observed in organic electrochemistry.

Participation of adsorbed intermediates can also be shown by the prolonged decay of the potential on interruption of the current (Conway and Vijh, 1967a) or by measurement of the time-dependence of the formation of products by carrying out the reaction with pulses of potential of controlled duration (Fleischmann et al., 1965). Thus the formation of ethane in the Kolbe reaction of acetate ions in acid solutions is initially proportional to the square of time as would be predicted for the rate $k\theta^2$ of the step (27) (Fleischmann et al., 1965).

$$CH_{3\,ADS} + CH_{3\,ADS} \rightarrow C_2H_6 \tag{27}$$

Stabilization of intermediates by strong adsorption will frequently be a necessary precondition for synthesis. Thus, in the case of the Kolbe reaction, further oxidation of the radicals is prevented: the formation of metal–carbon bonds in the reduction of alkyl halides (Fleischmann et al., 1971a; Galli and Olivani, 1970) or oxidation of Grignard reagents (Fleischmann et al., 1972c) is shown by the isolation of organometallic compounds. It is found that adsorption of the radical leads to a pronounced acceleration of the reaction and this can be readily understood in terms of the schematic potential energy–reaction coordinate plots (Fig. 5). Intervention of the adsorbed intermediate lowers the energy of activation in a manner exactly analogous to heterogeneous catalyses: the mechanism will in general depend on the free energy of adsorption and thereby on the metal (Parsons, 1958, 1968).

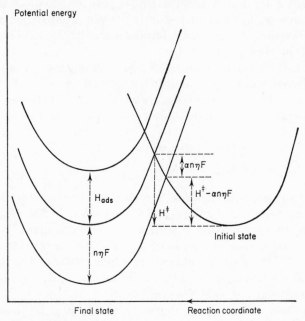

FIG. 5. Potential energy–reaction coordinate diagram for an electron transfer reaction leading to a product adsorbed on the electrode surface.

Although the concept of adsorption of organic species at the electrode surface is necessary to the understanding of many electrode processes, it will undoubtedly prove to be of special interest to the synthetic organic chemist since the adsorbed molecules are likely to have a fixed orientation with respect to the electrode surface and this gives rise to the possibility of stereospecific reactions (Feokstistov, 1968; Horner and Degner, 1971) and asymmetric syntheses (Horner and Skaletz, 1971; Horner, 1970) which are not possible to homogeneous solution. Stereospecificity in an electrode process is not a proof of adsorbed intermediates, however, since it can arise from free intermediates in solution by mechanisms involving hydrogen bonds or ion pairing as in homogeneous chemical reactions. Since the degree of adsorption and also the orientation of the adsorbate may be potential-dependent, it is clearly important to control the electrode potential in any investigation of the stereochemistry of electrode processes.

It has been known for many years that there are differences in the half-wave potentials for the reduction of geometric isomers, both when the isomerism is due to the presence of a double bond or a cyclohexane ring in the molecule. Typical is the case of maleic and fumaric acids where under certain conditions the *cis* isomer is more readily reduced by more than 100 mV (Kanakam *et al.*, 1967). These differences between the half-wave potentials for *cis* and *trans* isomers may often be correlated with differences in their adsorption isotherms (Korchinskii, 1968). More recently it has been shown that in reactions which produce unsaturated products, it is common to find stereospecificity and a variation in the *cis/trans* ratio in the product with changes in the electrolysis conditions. Thus the reduction of benzil produces the two isomers of stilbenediol and the *cis/trans* ratio has been shown to depend on the solvent, pH, and the electrode potential (Grabowski *et al.*, 1968; Vincenz-Chodkowska and Grabowski, 1964), while the oxidation of anthracene in acetonitrile containing methanol or acetate ion has been shown to yield a mixture of the *cis*- and *trans*-9,10-disubstituted-9,10-dihydroanthracenes, where the *trans* isomer is highly favoured (Parker, 1970).

The study of optical isomers has shown a similar development. First it was shown that the reduction potentials of several *meso* and racemic isomers were different (Elving *et al.*, 1955; Feokstistov, 1968; Závada *et al.*, 1963) and later, studies have been made of the ratio of *dl/meso* compound isolated from electrolyses which form products capable of showing optical activity. Thus the conformation of the products from the pinacolization of ketones, the reduction of double bonds, the reduction of onium ions and the oxidation of carboxylic acids have been reported by several workers (reviewed by Feokstistov, 1968). Unfortunately, in many of these studies the electrolysis conditions were not controlled and it is therefore too early to draw definite conclusions about the stereochemistry of electrode processes and the possibilities for asymmetric syntheses.

C. *The Electrode Surface*

Finally, the electrode potential may affect the overall process by determining the state of oxidation of the electrode surface. It is well known that in aqueous solution a platinum electrode has a bare surface only over the narrow potential range from approximately $+0.4$ V to $+0.8$ V versus N.H.E.: at more cathodic potentials it is covered by adsorbed hydrogen atoms while at more anodic potentials it is covered by adsorbed oxygen atoms or a layer of platinum oxide (Delahay, 1966). These changes in the electrode surfaces are reflected by the processes which take place. Oxidations which depend on the dissociative adsorption of the substrate occur only at potentials below those where the surface is covered by platinum oxide, and $i-E$ curves for such processes show a very sharp fall in current at the potential where the platinum oxide is formed. On the platinum oxide surface the reactions occur via electron transfer from an undissociated adsorbed or solute molecule. Although certain substrates may be oxidized on both types of surface, the mechanisms reflect the change in surface. Thus, acetic acid undergoes a slow catalytic oxidation to carbon dioxide on bare platinum (Koch and Woods, 1968; Woods, 1967) but an electron transfer-initiated process, the Kolbe reaction, on the oxide surface (Conway and Vijh, 1967b; Eberson, 1968). This does not occur until very positive potentials, i.e. $+2 \cdot 5$ V.

These changes in the state of oxidation of electrode surfaces and their attendant changes in reaction rates and mechanisms are very general for solid metal electrodes. For example, the surface of a nickel anode oxidizes in two steps; firstly, the nickel undergoes an irreversible oxidation to form a passive layer of nickel(II) hydroxide and then, just prior to oxygen evolution, a nickel(II) hydroxide $\rightarrow$ nickel(III) oxide transition occurs. It is found that a wide range of aliphatic amines and alcohols oxidize at the potential where the nickel(II) $\rightarrow$ nickel(III) change occurs and, indeed, it may be shown that the electrode reactions occur by the mechanism (Fleischmann et al., 1972a, 1971b).

$$Ni(OH)_2 \rightleftarrows NiO(OH) + H^+ + e$$

$$NiO(OH) + \text{organic molecule} \xrightarrow{\text{r.d.s.}} Ni(OH)_2 + \text{intermediate}$$

$$\text{intermediate} - ne \rightarrow \text{product.}$$

The oxidation of hydrazine follows the change in surface completely since it oxidizes rapidly on bare nickel and again on the nickel(III) oxide surface but in the intermediate potential region, where the surface is covered with nickel(II) hydroxide, the anodic oxidation cannot occur (Fleischmann et al., 1972d).

III. The Solution Environment

In this section it is intended to discuss the role of the solvent, the base electrolyte and the other reagents which are themselves not electroactive but which are added to vary the pH of the medium, to trap reaction intermediates or to vary the activity of the substrate, an intermediate or the product. It would seem correct, however, to discuss the various roles of the total environment of the reaction rather than to consider separately the individual components of the medium, since there is clearly often interaction between the components (e.g. an additive may be an acid in one solvent yet a base in another and the adsorption of electrolytes is known to be solvent-dependent) and it is the combination of the components which is seen by the electrode reaction.

Until about twenty years ago most organic electrochemistry was carried out in aqueous solution or in an alcohol or a carboxylic acid as the solvent, but the period since this time has seen a major expansion in the types of solvent which are considered suitable for electrochemistry. Firstly, there was a movement towards the study of electrode reactions in aprotic solvents such as acetonitrile, dimethylformamide and dimethyl-sulphoxide since it was found that electron transfer reactions were generally faster and reaction intermediates more stable in these solvents than in protonic media. More recently, studies have been extended to a more ambitious range of media, including poorly conducting solvents [e.g., diethyl ether (Gillet, 1961)], strongly acidic media [e.g., fluorosulphuric acid (Bertram *et al.*, 1971)], strongly basic media [e.g., HMPA (Sternberg *et al.*, 1968)] and to compounds which are not liquids at N.T.P., namely condensed gases [e.g. sulphur dioxide (Miller and Mayeda, 1970)] and molten salts (Sundermeyer, 1965)]. There has been a consequent broadening in the type of electrode reactions which can be observed. There has been a similar development in the types of base electrolyte which are used and the introduction of tetraalkylammonium salts was of special importance (Mann, 1969).

The functions of the solution environment will be considered under four sub-headings which are basic requirements, the environment as a reactant, pH effects and double layer and adsorption effects.

A. *Basic Requirements*

It is normally necessary for the solvent to dissolve both ionic and organic species. The ionic species are required so that the media are conducting. Commonly, the base electrolyte is an inorganic salt, acid or base or a tetraalkylammonium salt but it may also be formed by interaction of an organic or organometallic compound with the solvent. Examples of the latter cases are acetic acid in fluorosulphuric acid (Bertram *et al.*, 1971) and Grignard reagents in ethers (Braithwaite, 1969a). Furthermore, modern potentiostats allow systematic studies in very poorly conducting media although large-scale electrolyses require higher conductivity. Even this restriction may be removed if modern, unconventional cell designs are used (see Section IX).

The other major basic requirement is normally that the environment is electrochemically inert at the potential at which the substrate undergoes oxidation or reduction. A long cathodic range is obtainable in many solvents if tetraalkylammonium salts are used as the base electrolyte or the presence of free hydrogen ion and cathodes of low hydrogen overpotential are avoided. On the anodic side, long potential ranges are only found in a few solvents. In aqueous solution, oxygen evolution is retarded in acid solution and by anion adsorption, and electrolyses are possible to $+3\cdot0$ V. Similar potential ranges are possible in certain aprotic solvents, namely acetonitrile, nitromethane, propylene carbonate, dichloromethane and sulphur dioxide [if a tetrafluoroborate or a hexafluorophosphate is used as the electrolyte (Fleischmann and Pletcher, 1968)] and also in hydrogen fluoride (Doughty et al., 1972).

B. *The Environment as a Reactant*

1. *Reaction with intermediates*

Perhaps the most important single function of the solution environment is to control the mode of decomposition of reaction intermediates and hence the final products. This is particularly true in the case of electrode reactions producing carbonium ion intermediates since the major products normally arise from their reaction with the solvent. It is, however, possible to modify the product by carrying out the electrolysis in the presence of a species which is a stronger nucleophile than the solvent and, in certain non-nucleophilic solvents, products may be formed by loss of a proton or attack by the intermediate on further starting material if it is unsaturated. The major reactions of carbonium ions are summarized in Fig. 6.

In the case of carbanion and radical intermediates the solvent is less important but the products are partially determined by the resistance of the medium to proton or hydrogen atom abstraction respectively. The increased stability of these intermediates compared with carbonium ions allows the reaction mechanism to be more readily modified by the addition of trapping agents. For example, carbanions are trapped in high yields by the presence of carbon dioxide in the electrolysis medium (Wawzonek and Wearring, 1959; Wawzonek et al., 1955).

The role of organic intermediates in electrode reactions was recently reviewed in some detail (Fleischmann and Pletcher, 1969 and 1971).

2. *Stabilization of intermediates*

Certain solvents stabilize intermediates by strong solvation. The best known example is that of the electron itself which forms stable solutions in a number of solvents including liquid ammonia and hexamethylphosphoramide. These solutions are extremely strong reducing agents and also highly basic media. The solvated electron is formed at highly cathodic potentials where it passes from the electrode directly into the solvent. It is apparently a better reducing agent than its counterpart at the electrode surface, since it is able to reduce a number of compounds, for example, benzene (Benkeser *et al.*, 1964; Dubois and Dodin, 1969; Sternberg *et al.*, 1963, 1966, 1969), amides (Avaca and Bewick, 1971; Benkeser *et al.*, 1970), and nitriles (Arapakos and Scott, 1968) which are themselves electrochemically inactive in aprotic media. This is probably because of easier electron orbital overlap in the case of the solvated

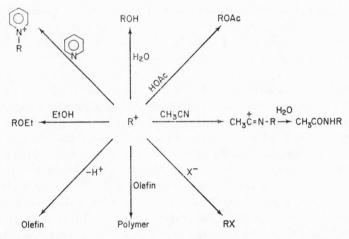

FIG. 6. Typical reactions of simple carbonium ions in various solution environments.

species. The solvated electron also has advantages when scaling up a process in that it is possible to reduce dilute solutions of an organic substrate at unusually high current densities. Since the electron diffuses at a higher rate than any organic species, the electron is able to diffuse into the bulk of the solution much faster than the substrate can diffuse to the electrode surface. The reduction therefore occurs at the outer boundary of the diffusion layer and it is possible to exceed the normal diffusion-controlled current by several orders of magnitude without loss of current efficiency (Avaca and Bewick, 1971).

Carbonium ions are likewise known to be stabilized by solvation in strongly acidic media (Olah and Pittman, 1966), and consequently the reduction potential of carbonium ions, the ease of formation of a carbonium ion by the oxidation of a substrate and the products from these reactions would be expected to depend on the acidity of the electrolysis medium. Detailed studies comparing the solvating ability of solvents for organic ions have not been reported. Such studies are available, however, for simple inorganic ions and it has been shown, using formal electrode potentials obtained from polarography, that the solvation energies for metal cations decrease in the solvent series $DMSO > DMF \simeq H_2O >$ propylene carbonate $> CH_3CN > CH_3NO_2 > C_6H_5NO_2$ (Pletcher, 1967). This information can be used to achieve the stabilization of unusual oxidation states. It is well known that ions of higher valency are stabilized by strongly solvating or complexing media, and conversely poorly solvating solvents favour the stabilization of lower oxidation states. For example, In^+ and Sm^{++} are stable species in acetonitrile and nitromethane but not in water (Headridge and Pletcher, 1967; Kolthoff and Coetzee, 1957). Similarly the strong affinity of Ag^+ and Cu^+ for nitrogen ligands is reflected by their great stability in acetonitrile (Coetzee et al., 1963).

Intermediates in electrode reactions are also stabilized by ion pair formation, particularly in organic solvents with relatively low dielectric constants. Thus the reduction potentials of the radical anions of quinones in dimethylformamide are markedly affected by the cation of the inert electrolyte owing to interaction of the cation with the product of the reduction process, the dianion (Peover and Davies, 1963). The half-wave potential becomes less cathodic along the series Bu_4N^+, Et_4N^+, K^+, Na^+, Li^+, H^+, as would be predicted by consideration of their cationic radii; the half-wave potential for the anthraquinone ion-radical is over 0.5 V more positive in the lithium electrolyte than in a tetrabutylammonium salt, and ion pair formation with lithium ion has also made the process more reversible. The degree of complex formation is, however, partly dependent on the degree of charge delocalization in the anion and also on the solvation of the cation. Thus, the dependence of the degree of ion pairing on the cations is different for large, delocalized aromatic anions in ether solvents (Bhattacharyya et al., 1965; Buschow et al., 1965). Here it is found that the tetraalkylammonium ion forms stronger ion pairs than lithium ions, and this may be attributed to the fact that in these solvents the tetraalkylammonium ion is poorly solvated and hence smaller than the solvated lithium ion. There is also a dependence on the polarity of the organic anion since for p-benzoquinone the order reverts to $Li^+ > Et_4N^+$ (Peover, 1967).

3. Reaction of the medium at the electrode

Certain electrode reactions occur by a mechanism which involves initial oxidation or reduction of one of the components of the electrolysis medium. This species is commonly reformed later in the reaction sequence and, hence has the role of a catalyst. Typical examples of this type of reaction are the anodic oxidation of propylene to propylene oxide in presence of aqueous chloride ion (Ibl and Selvig, 1970),

$$Cl^- + H_2O \xrightarrow{-2e-2H^+} ClO^-$$

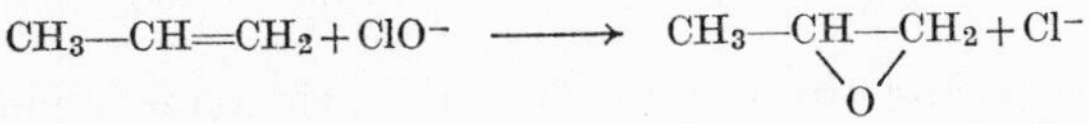

$$CH_3\text{—}CH\text{=}CH_2 + ClO^- \longrightarrow CH_3\text{—}CH\text{—}CH_2 + Cl^-$$

the formation of heterocycles during the oxidation of acetonitrile containing perchlorate ion (Fleischmann and Pletcher, 1972),

$$ClO_4^- \xrightarrow{-e} ClO_4^{\cdot}$$

$$ClO_4^{\cdot} + CH_3CN \xrightarrow{CH_3CN} \text{heterocycles} + HClO_4$$

and the oxidation of toluene in methanol containing methoxide ion (Tsutsumi and Koyama, 1968),

$$MeO^- \xrightarrow{-e} MeO^{\cdot}$$

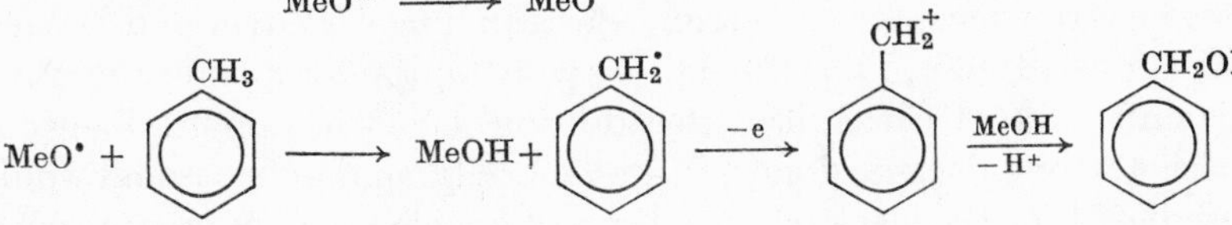

The evidence for this type of mechanism is usually that the steady-state i–E curve for the electrolysis medium is unaltered by the addition of the substrate.

4. *Reaction of the environment with the starting material*

The commonest example of this type of interaction is the protonation of the substrate by acids in the electrolysis medium, but pH effects will be dealt with in a later section. There are, however, other chemical interactions which can occur. For example, the mechanism and products of the oxidation of olefins are changed by the addition of mercuric ion to the electrolysis medium. In its absence, propylene is oxidized to the allyl cation (Clark et al., 1972),

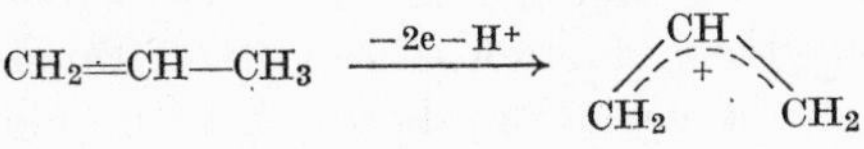

$$CH_2\text{=}CH\text{—}CH_3 \xrightarrow{-2e-H^+}$$

but in its presence, the reaction sequence becomes (Fleischmann $et\ al.$, 1969, 1970),

$$CH_3—CH{=}CH_2 \xrightarrow[-H^+]{Hg^{2+}+H_2O} \underset{\substack{|\quad\ \ | \\ OH\ \ Hg^+}}{CH_3—CH—CH_2} \xrightarrow{-2e} Hg^{2+}+\underset{\substack{| \\ OH \\ \downarrow \\ products}}{CH_3—CH—CH_2^+}$$

and it may be noted that, again, the mercuric ion is unchanged by the overall process.

Weaker complex formation may affect the potential at which the electrode reaction will occur but without changing the products. The complex formation may be ion pairing or the formation of charge transfer complexes. An example of the former type of interaction is the oxidation of acetate ion in acetonitrile where the potential at which the Kolbe reaction occurs is very dependent on the cation of the inert electrolyte [Et_4N^+; $+1\cdot2$ V, K^+; $+1\cdot6$ V and H^+; $+2\cdot9$ V versus a Ag/Ag^+ reference electrode (Fleischmann and Pletcher, 1972)]. Charge transfer complexes are formed by pairs of hydrocarbons in solvents of low dielectric constant and the half-wave potential of aromatic hydrocarbons have been reported to be affected by the addition of a second hydrocarbon which is either a strong electron acceptor or a strong electron donor (Peover and Davies, 1964; Peover, 1967). Thus the half-wave potential for tetracyanoethylene is shifted 39 mV cathodic by the addition of $0\cdot5$ M hexamethylbenzene which is a strong electron donor. More generally, similar behaviour would be predicted for the electrode reactions of a wide range of organic compounds which are capable of acting as a Lewis acid or a Lewis base and hence can form charge transfer complexes by acid-base interaction.

C. pH Effects

The role of the pH of the medium in the electrode reactions of organic compounds in aqueous solutions is well understood and has been recently reviewed in detail (Zuman, 1969). In particular, our understanding of this parameter is due to the large number of polarographic investigations where it has been found that the half-wave potential, the limiting current and the shape of the wave for an oxidation or reduction process may all be dependent on the acidity of the medium.

The absence of any variation of these characteristics of the electrode reaction with pH indicates that it is the species which predominates in the bulk of the solution which undergoes the electron transfer at the electrode surface. Conversely, a variation shows that the electrode process is distorting an equilibrium which exists in the bulk of the solution. If one considers the acid-base equilibrium

$$A \underset{k_2}{\overset{k_1}{\rightleftarrows}} B + mH^+$$

where only the acid or the base form of organic species is electroactive at the potential of the dropping mercury electrode, i.e.,

$$A \pm ne \rightarrow P_1$$

or

$$B \pm ne \rightarrow P_2$$

two types of behaviour are commonly observed.

(a) $E_{1/2}$ but not i_L, the limiting current, is pH dependent. Here both forms of the organic species are transported to the electrode surface, the equilibrium is rapid so that the overall process is diffusion-controlled and the transfer of m protons precedes the electron transfer. A plot of $E_{1/2}$ versus pH should be linear and its slope is $\pm mRT/\alpha nF$: it therefore gives information about m since $\pm RT/\alpha nF$ may be obtained from an analysis of the shape of the polarographic wave (Meites, 1965; Zuman, 1969).

(b) Both $E_{1/2}$ and i_L are dependent on pH. In this case the equilibrium is slow so that the overall process is controlled kinetically by the rate of the preceding proton transfer. It is possible that a second wave is observed, at more negative potentials in the case of reductions, which is due to an electron-transfer process of the form of the organic compound which was not electroactive at the lower potential. Again m may be determined from a $E_{1/2}$–pH plot.

At the extremes of pH it is common for the equilibrium to be driven completely to the left or right and then the electrode process becomes independent of pH since it is the bulk species which is electroactive. Thus the common shape of pH–$E_{1/2}$ curves is shown in Fig. 7 (Zuman and Tang, 1963).

The principles outlined above are, of course, important in electrosynthetic reactions. The pH of the electrolysis medium, however, also affects the occurrence and rate of proton transfers which follow the primary electron transfer and hence determine the stability of electrode intermediates to chemical reactions of further oxidation or reduction. These factors are well illustrated by the reduction at a mercury cathode of aryl alkyl ketones (Zuman et al., 1968). In acidic solution the ketone is protonated and reduces readily to a radical which may be reduced further only at more negative potentials.

$$ArCO.R + H^+ \rightleftharpoons ArC^+(OH)R$$

$$ArC^+(OH)R + e \xrightleftharpoons{E_1} ArC^\cdot(OH)R \longrightarrow \begin{array}{c} ArCR\text{---}CRAr \\ | \quad | \\ OH \quad OH \end{array}$$

$$ArC^\cdot(OH)R + e \xrightarrow{E_2} ArC^-(OH)R \xrightarrow{H^+} ArCH(OH)R$$

Hence two one-electron waves are observed on a polarogram and two products, either the pinacol or the alcohol, may be isolated depending

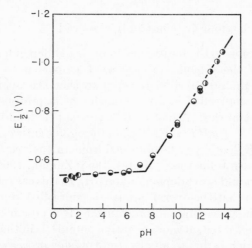

Fig. 7. Effect of pH on the half-wave potential of methyl butyl phenacyl sulphonium perchlorate [2×10^{-4} M] in water + 0.2% ethanol. ◑ Britton-Robinson buffers, ◐ sulphuric acid, ◐ sodium hydroxide. (Taken from Zuman and Tang, 1963.)

on the potential of the electrolysis. In the medium pH range the ketone remains unprotonated and hence is reduced less readily than in acid solution. The anion radical which is formed initially is, however, protonated and the radical formed reduces more readily than the parent ketone.

$$ArCO.R + e \xrightarrow{E_1} ArC^\cdot\text{-}OH$$

$$ArC^\cdot\text{-}OR + H^+ \rightleftharpoons Ar\dot{C}(OH)R$$

$$Ar\dot{C}(OH)R + e \xrightarrow{E_1} ArC^-(OH)R \xrightarrow{H^+} ArCH(OH)R$$

Thus a single two-electron wave is observed and only one product, the alcohol, can be isolated. Finally, at high pH neither the ketone nor the radical anion are protonated by this basic medium and it is not until the dianion, formed by successive electron transfers, that protonation occurs.

$$ArCO.R + e \xrightarrow{E_1} ArC^{\cdot-}OR \longrightarrow \underset{\underset{O^-\;O^-}{|\quad|}}{ArCR-CRAr} \xrightarrow{2H^+} \underset{\underset{OH\;OH}{|\quad|}}{ArCR-CRAr}$$

$$Ar\overset{\cdot}{C}OR + e \xrightarrow{E_2} Ar\overset{-}{C}-\overset{-}{O}R \xrightarrow{2H^+} ArCH(OH)R$$

Again two one-electron polarographic waves are obtained and two different products may be isolated.

More recently it has become apparent that proton equilibria and hence "pH" can be equally important in aprotic and other non-aqueous solvents. For example, the addition of a proton donor, such as phenol or water, to dimethylformamide has a marked effect on the i–E curve for the reduction of a polynuclear aromatic hydrocarbon (Peover, 1967). In the absence of a proton donor the curve shows two one-electron reduction waves. The first electron addition is reversible and leads to the formation of the anion radical while the second wave is irreversible owing to rapid abstraction of protons from the solvent by the dicarbanion.

$$RH \underset{E_1}{\overset{+e}{\rightleftharpoons}} RH\cdot^- \xrightarrow[E_2]{+e} RH^{--} \xrightarrow{solvent} RH_3$$

After addition of an excess of a proton donor, a single two-electron wave is observed. This is because the radical anion protonates to give a radical which is more readily reduced than the parent hydrocarbon

$$RH \xrightarrow[E_1]{+e} RH\cdot^- \xrightarrow{H^+} RH_2 \xrightarrow[E_1]{+e} RH_2^- \xrightarrow{H^+} RH_3$$

One of the attractions of aprotic solvents is that the electron transfer behaviour of many compounds is much simpler than in protonic media. However, this is not always so; for example, the quinone/hydroquinone couple is very simple in aqueous solution but it is complicated in aprotic solvents by the number of protonation equilibria which no longer lie well to one side as they do in aqueous solution (Bessard $et\ al.$, 1970).

A further difficulty arises during preparative electrolyses in aprotic solvents because of the bulk "pH" change which commonly occurs. Thus cathodic reductions often require proton abstraction from the solvent in order to yield stable products, while many anodic oxidations, including those of aromatic and aliphatic hydrocarbons, give rise to a quantitative yield of proton and the consequent changes in the "pH" of the electrolysis media would be expected to lead to a variation in the products with the duration of the electrolysis. Unfortunately, the "pH" can be a very difficult parameter to control in aprotic solvents and most work reported in the literature has been carried out in unbuffered conditions. In the case of oxidations, organic bases, e.g. pyridine, have

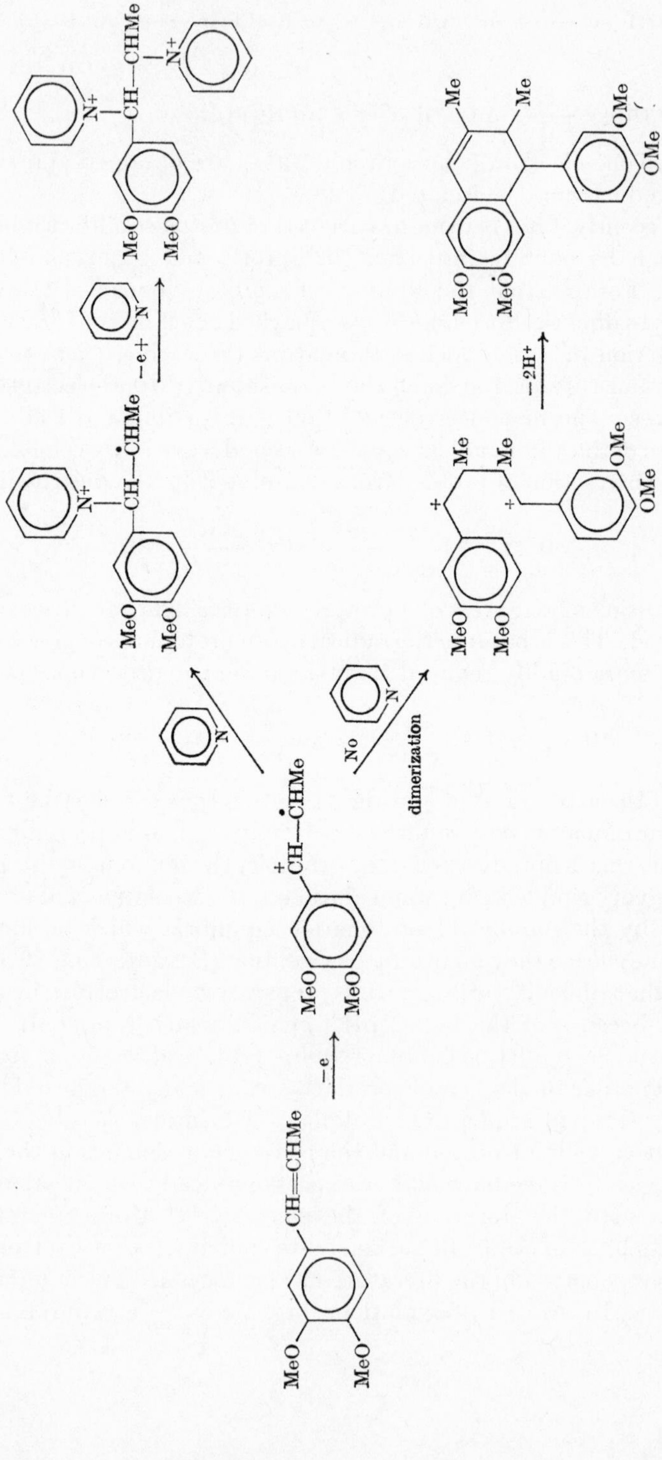
−e
No
dimerization
−e +
−2H+

been used in an attempt to maintain a constant "pH". However, it was found that entirely different products were often isolated from electrolyses after the addition of pyridine. For example the oxidation of 3,4-dimethoxypropenylbenzene had the reaction routes shown on page 182 (O'Connor and Pearl, 1964). Later Eberson and Parker (1969) emphasized that many species in the electrolysis medium have the dual roles of base and nucleophile. They used a cyclic voltammetric technique to measure the rate of reaction between three lutidines and three cation radicals; the results are shown in Table 1. The cation radicals

TABLE 1

Relative Rates of Reactions Between Cation Radicals and Lutidines

Cation radical from	3,5-lutidine	2,5-lutidine	2,6-lutidine
9,10-Diphenylanthracene	1·00	0·15	0·03
1,4-Dimethoxybenzene	0·48	0·03	0·02
9,10-Dimethylanthracene	1·34	1·00	0·80

have the choice of two reactions with the lutidine which can act as a base or a nucleophile

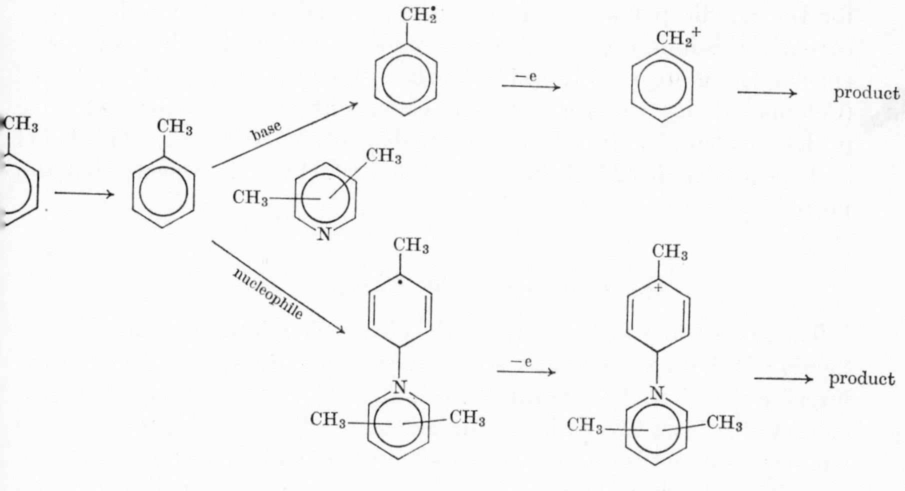

Where the three lutidines behave primarily as bases there is no marked difference in their reaction rates but where they react as nucleophiles,

7

steric factors around the nitrogen atom are important and the α,α' disubstituted compound reacts appreciably more slowly than the less hindered lutidines. Thus, in their reactions with the cation radical of 9,10-dimethylanthracene their role is that of a base, while in the other cases they act as nucleophiles. Clearly, anions as well as organic bases have these dual roles and, for example, acetate and fluoride ion may be expected to be reasonably strong bases (Parker, 1969). Recently, there have been attempts to use insoluble inorganic bases, e.g. CaH_2 or $NaHCO_3$, to maintain a constant "pH" in the bulk of the solution, but these compounds clearly cannot buffer the electrode reaction itself (Fleischmann and Pletcher, 1972; Miller *et al.*, 1971).

During the last few years it has been realized that there are interesting possibilities for electrosynthetic reactions over a wider range of acidity than may be obtained in aqueous or aprotic solvents. The extremely basic solvents, e.g. methanol/methoxide ion or hexamethylphosphoramide/solvated electron, and the strongly acidic solvents, e.g. liquid hydrogen fluoride, fluorosulphonic acid or benzene/aluminium trichloride/hydrogen chloride, offer possibilities for the activation of inert substrates. The highly acidic media are particularly useful media for the oxidation of hydrocarbons; fluorosulphuric acid containing acetic acid has been shown to be a solvent for the oxidation of alkanes to $\alpha\beta$-unsaturated ketones (Bertram *et al.*, 1971) and the liquid complex benzene/aluminium trichloride/hydrogen chloride has been used as a solvent for the anodic polymerization of benzene and other aromatic hydrocarbons (Wisdom, 1969). There is now good evidence that the protonated species are more readily oxidized than the hydrocarbons themselves (Coleman *et al.*, 1972) and this is consistent with chemical evidence (Coleman *et al.*, 1972; Kollmar and Smith, 1970; Olah *et al.*, 1967, 1971) and spectroscopic data (Olah *et al.*, 1967, 1971) which indicate that the protonated form is more reactive.

D. *Double Layer and Adsorption Effects*

The expressions for the rates of the electrochemical reactions given in Section II.A have not taken into account the detailed structure of the interfacial region. In general, the solution adjacent to the electrode will consist of at least two regions. Immediately adjacent to the metal there will be a compact layer of ions and solvent molecules which behaves as a capacitor. A potential difference $(\phi_m - \phi_2)$ will be established between the metal and the plane of closest approach, designated the ϕ_2 plane. In the case of desolvation and specific adsorption a further ϕ_1 plane can be designated (Fig. 8) but this will not be considered further here.

Between the ϕ_2 plane and the body of the solution a diffuse space charge layer establishes the potential ϕ_2 and the magnitude of this potential depends on ϕ_m and the ionic strength of the solution. It will be apparent that ϕ_2 will determine the concentrations of charged electroactive species, while $(\phi_m - \phi_2)$ will determine the rate of the electron transfer step if electron transfer takes place at the plane of closest approach. This leads

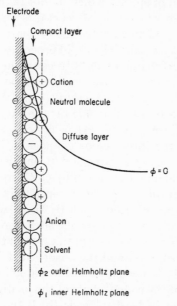

FIG. 8. Schematic representation of the electrode-solution interface and the potential distribution in this zone.

to a modification of equation (10). For example, for the cathodic discharge of a species carrying a charge $+z$

$$i = i_{0,\text{true}} \exp\left(\frac{(\alpha n - z)F\phi_2}{RT}\right)\left[\exp\left(\frac{-\alpha nF}{RT}\eta\right) - \exp\left(\frac{(1-\alpha)nF}{RT}\eta\right)\right] \quad (28)$$

so that

$$i_0 = i_{0,\text{true}} \exp\left(\frac{(\alpha n - z)\phi_2 F}{RT}\right) \quad (29)$$

Equation (28) shows that changes in the structure of the interfacial region can lead to catalysis through purely physical factors, namely the distribution of potential (Frumkin, 1961). Thus, if the reactant is uncharged and a radical anion is generated, then a positive shift in ϕ_2 would lead to an increase in the rate of reaction. Marked effects of this kind have been observed for inorganic systems (Delahay, 1966) but have not been explored in organic electrosynthesis. It should be noted that ϕ_2 will change with electrolyte concentration and, in particular, with specific adsorption of ions.

It is unlikely that the simple effect outlined above will be observed in isolation from other changes. In the first place, the ϕ_2 potential may control the concentration of species which do not take part in the reaction but in turn affect the adsorption of reactants and intermediates; the rates of homogeneous chemical reactions of charged intermediates will also be affected. These are normally rapid processes which will frequently take place in the space charge region. In consequence, the rates will be affected both by changes in the concentration distribution of the intermediates and other charged reactants (e.g. H_3O^+) and by the effects of the high fields (up to 10^7 V cm^{-1}) on the rates (field dissociation) (Nürnberg, 1965). Furthermore, other factors, such as changes in solvent structure in the interface or changes in ion pairing, will also exert a strong influence on the rates and mechanisms.

At this stage of development of the subject it is appropriate to consider a number of empirical rules which may serve to indicate the important variables. It would seem likely that (i) there will be competitition between each species in the system for the sites available at the electrode surface, and that (ii) for each species in the system there will be an equilibrium between the solution and the adsorbed state. Thus it would be expected that the solution constituents would affect these equilibria in two ways: (a) if one of the constituents of the medium is itself adsorbed, the reactant will tend to be displaced; (b) if the reactant is strongly solvated, complexed or ion paired by constituents of the medium, the species in solution will be favoured.

Unfortunately, the only detailed and systematic studies of the effect of the solution environment on adsorption have been carried out at a mercury electrode and with species which are uninteresting from the point of view of electrosynthesis. Generally, these studies using electrocapillary measurements and capacitance–potential curves have, however, supported the above hypotheses. For example, it has been shown that for series of butyl, phenyl and naphthyl derivatives there is an inverse linear correlation between their standard free energies of adsorption and their standard free energies of solvation (Blomgren et al., 1961). The adsorption of ionic species is also a reflection of solution properties. Hence, the coverage of the electrode by ions increases along the solvent series, dimethylsulphoxide, water, propylene carbonate, a reflection of the fact that propylene carbonate solvates ions less than water, etc. (Reeves, 1969). Similarly, anions are adsorbed more strongly than ca-

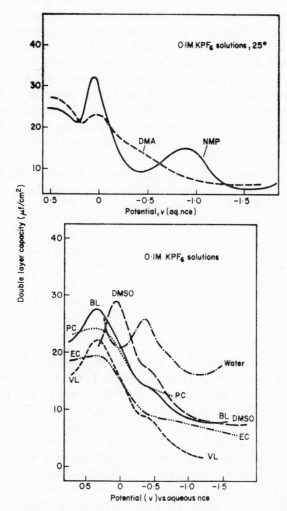

Fig. 9. Capacitance–potential curves for a number of common electrochemical solvents containing 0·1M potassium hexafluorophosphate, a relatively non-adsorbing electrolyte. (From Payne 1967, 1970.)

tions from water under similar conditions because of their lower solvation energies, while tetraalkylammonium cations are adsorbed less readily from aprotic solvents than from water owing to their higher free energies of solvation in these solvents (Peover and Dietz, 1968).

The solvents themselves are adsorbed on the electrode surface, as is shown by the capacitance–potential graphs illustrated in Fig. 9 (Payne, 1967, 1970); potassium hexafluorophosphate, the electrolyte in each of the solvents, is thought to be adsorbed only very weakly. The solvents show somewhat differing curves and the peaks have been interpreted both in terms of competition between the solvent and anions for sites at the surface and also in terms of solvent reorientation. Ethers are adsorbed from the amide solvents most strongly at the potentials around the peaks and this has been postulated to be due to an increase in freedom for the solvent to rotate at these potentials (Dutkiewicz and Parsons, 1966).

Perhaps the best known example of adsorption effects in electrosynthetic reactions is the beneficial role of tetraalkylammonium ions in the hydrodimerization of acrylonitrile

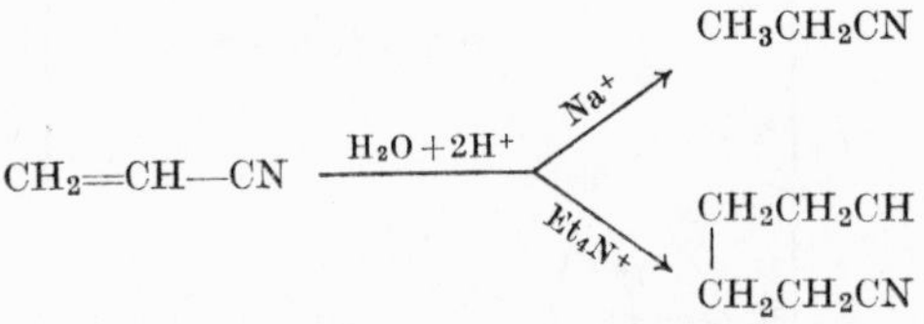

$$
CH_2{=}CH{-}CN \xrightarrow{\ H_2O+2H^+\ }
\begin{array}{l}
\overset{Na^+}{\nearrow} \quad CH_3CH_2CN \\[2mm]
\underset{Et_4N^+}{\searrow} \quad
\begin{array}{l}
CH_2CH_2CH \\
| \\
CH_2CH_2CN
\end{array}
\end{array}
$$

where the presence of these ions changes the yield of adiponitrile from zero to almost quantitative (Baizer, 1964). It has been suggested that the function of the onium salt is to provide an anhydrous layer at the electrode surface which favours adsorption of the organic compound but which repels water (Gillet, 1968). The fact that the "inert" ions participate in the electrode reactions is also clearly shown, for example, by the variation of the dl/meso ratio of the pinacol formed from the reduction of isobutyrophenone with the nature of the tetraalkylammonium cation (see Table 2) (Horner and Degner, 1971). In the presence of optically active cations the extent of asymmetric induction and absolute configuration of the product depend on the configuration and structure of the cation (Horner and Degner, 1971; Horner and Skaletz, 1971). This shows that the reactant molecules must be held in a preferred orientation with respect to the chiral salt in the double layer and coadsorption of species must in general be considered.

TABLE 2

Percentage Yields of *meso* and D,L Product Formed During the
Pinacolization of Isobutylphenone in the Presence of Various
Base Electrolytes

Base electrolyte	Isobutyl phenyl pinacol	
	meso	D,L-
$(CH_3)_4N^+Cl^-$	84	16
C_6H_5—C—C—CH$_3$ (H H, OH N$^+$(CH$_3$)$_3$Cl$^-$)	70	30
C_6H_5—C—C—CH$_3$ (H H, OH N$^+$H$_2$CH$_3$Cl$^-$)	5	95

A further effect which has been known for many years is that of anions, which are specifically adsorbed at high anodic potentials on platinum, on the products of the oxidation of carboxylate ions. For example, carbonium ion-derived products can be obtained in the presence of such specific adsorption and this demands a complete change in reaction route (Fioshin and Avrutskaya, 1967; Glasstone and Hickling, 1934).

At least two other unusual solvent effects have recently been reported and may be due to differences in adsorption equilibria. Ethyl bromide reduction in propylene carbonate has been reported to yield lead tetra-ethyl only in the presence of an onium ion (Galli and Olivani, 1970), while in dimethylformamide the organolead compound may be formed in very high yield by electrolysis with sodium bromide as the electrolyte (Fleischmann *et al.*, 1971a). Secondly, the product from the electrolysis of methylbenzenes in acetonitrile containing 1% water (see page 190) has been shown to depend on the anion of the base electrolyte (Eberson and Olafson, 1969). This later example has been challenged (Miller and Mayeda, 1972) since the benzyl alcohol can be shown to undergo an acid-catalyzed conversion to the amide in the presence of perchlorate and under the conditions which prevail during an unbuffered electrolysis. This conversion does not occur under the same conditions in the presence of the tetrafluoroborate anion as this is more basic than the perchlorate ion. Thus it would appear unnecessary to postulate that tetrafluoro-borate preferentially takes water into the electrode interface as was done in the original paper.

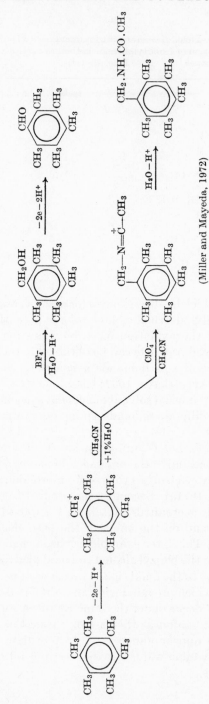

(Miller and Mayeda, 1972)

IV. Electrode Material

As with solvents and electrolytes, the past two decades have seen a considerable broadening in the range of materials which are considered suitable as electrodes. The popularity of the dropping mercury electrode with analytical, physical organic and theoretical electrochemists may be ascribed to its highly reproducible nature; since each drop has a lifetime of only a few seconds, one is assured of a new, clean surface at short, regular intervals and this leads to a simplicity in the resultant $i-E$ curves which is seldom found with solid electrodes. In fact, the advantages of the dropping mercury electrode have led to a reluctance to use other electrode materials and it was not until a real desire existed to scale up electrode reactions and to study reactions at potentials which were not accessible on mercury that real progress was made. Reductions have now been attempted using most metals as a cathode, although mercury, lead and tin remain the most popular cathode materials because of their high hydrogen overpotentials, while the platinum metals and carbon are the most widely used electrodes for anodic processes. Recently, however, a number of non-metallic materials have been investigated as electrodes for organic reactions. These include organic semiconductors, such as anthracene (Mehl and Hale, 1968) and metal oxides (Fleischmann et al., 1971b, c; Hampson et al., 1969, 1970; Horanyi et al., 1971), and a dimensionally stable ruthenium dioxide-covered titanium electrode, developed by de Nova (1970), has been shown to be a superior anode for chlorine evolution and certain organic oxidations.

All solid electrodes are liable to undergo surface changes which depend on the electrode potential and the solution conditions. For example, in aqueous solution, platinum electrodes have only a relatively narrow range where their surface is not covered with either a platinum oxide, absorbed oxygen atoms or adsorbed hydrogen atoms. Since some of these changes are irreversible, the performance of solid electrodes is dependent on their previous history and their method of preparation (Gilman, 1967): to quote an extreme example of the latter, there is no resemblance between the activity of platinum black and smooth platinum electrodes. In addition, the surface of solid electrodes is readily poisoned by a trace impurity in the electrolysis solution or by a minor electrode intermediate. Thus, clearly, extreme care and thought are required in the handling of solid electrodes, and several approaches to the problem of producing a clean, reproducible surface have been reported. These methods usually involve treatment with chemical oxidizing or reducing agents, placing the electrode in a hot flame or applying complex potential pulse profiles where the various potentials are designed to desorb or oxidize organic contaminates and to leave the electrode free of oxide or hydrogen atoms (Gilman, 1967). Furthermore, in analytical and preparative work there have been various ingenious attempts to produce a clean surface and to imitate the dropping mercury electrode. These involve scraped and wiped electrodes (Spencer, 1969) and the use of carbon as a paste (Kuwana and French, 1964), in the form of carbon felt (Bobbitt *et al.*, 1971) and even as a "carbon toothpaste electrode" where a renewed surface is produced by constantly injecting a stream of carbon paste into the preparative cell (Bobbitt, 1972).

Hence, it is important to remember that the products, reaction mechanism and the rate of the process may depend on the history and pretreatment of the electrode and that, indeed, the activity of the electrode may change during the timescale of a preparative electrolysis. Certainly, the mechanism and products may depend on the solution conditions and the electrode potential, purely because of the effect of these parameters on the state of the electrode surface.

Before considering the role of the electrode material in detail, there is one further factor which should be pointed out. The product of an electrode process may be dependent on the timescale of the contact between the electroactive species and the electrode surface, particularly when a chemical reaction is sandwiched between two electron transfers in the overall process. This was first realized when it was found that *i–E* curves and reaction products at a dropping mercury electrode were not always the same as those at a mercury pool electrode (Zuman, 1967a). For example, the reduction of *p*-diacetylbenzene at a mercury pool was found to be a four-electron process, giving rise to the dialcohol, while at a dropping mercury electrode the product was formed by a two-electron process where only one keto group was reduced (Kargin *et al.*, 1966). These facts were interpreted in terms of the mechanism

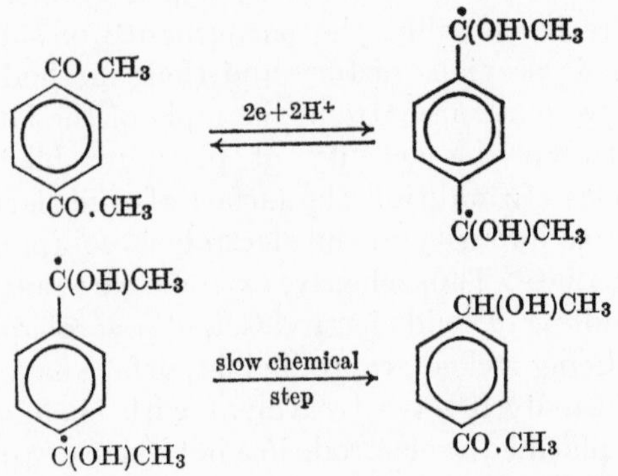

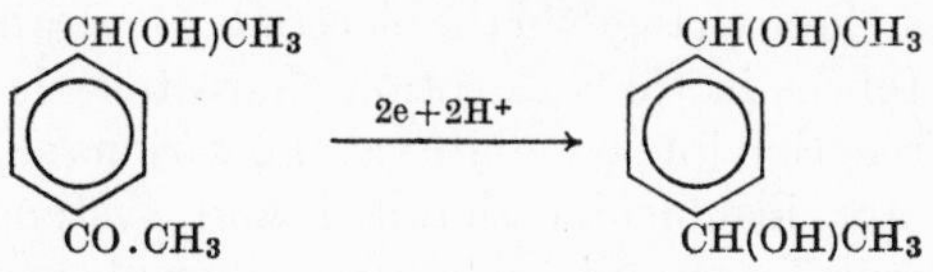

$$\text{(structure with CH(OH)CH}_3 \text{ and CO.CH}_3) \xrightarrow{2e+2H^+} \text{(structure with CH(OH)CH}_3 \text{ and CH(OH)CH}_3)$$

where the chemical reaction is slow and therefore does not occur during the lifetime of the mercury drop. Similar behaviour may be observed at a rotating disc electrode; here the product can depend on the rate of rotation of the disc. For example, the oxidation of 9,10-diphenylanthracene [equation (30)] at platinum and in the presence of 4-methylpyridine is a two-electron process at a stationary or slowly rotating electrode but a one-electron process at high rotation rates (Manning *et al.*, 1969).

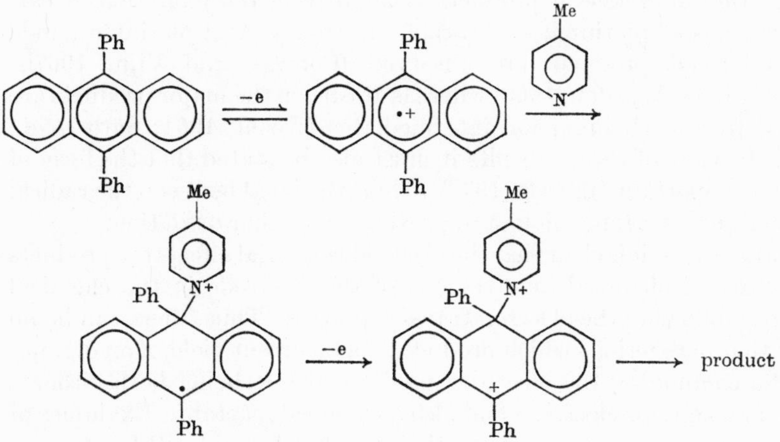

$$\text{(reaction scheme)} \qquad \text{product} \quad (30)$$

Thus, under the hydrodynamic conditions prevailing at high rotation rates, the one-electron product is removed more rapidly by convection than by the chemical reaction, while at slow rotation speeds the chemical reaction and further electron transfer predominates. The form of the electrode and the hydrodynamic conditions prevailing in the electrolysis solution are therefore parameters which require controlling but which give additional flexibility in the design of syntheses.

The electrode is often considered to be inert and its role is simply to act as a source or a sink for electrons, depending on its potential. It is clear, however, that the rate of many electrode processes, and indeed their products, is dependent on the material from which the electrode is constructed.

In certain cases, these changes in the electrode process may be explained in terms of differences in the adsorption characteristics of the electro-active species or reaction intermediates at the various electrode surfaces, although there are insufficient detailed and systematic studies of adsorption isotherms for organic substrates at a wide range of electrodes to test such hypotheses. As has been discussed earlier, some metals appear to catalyse the dissociative adsorption of organic substrates, and the reversible chemisorption of species would also be expected to depend on the crystallographic and electronic structure of the electrode material. Certainly in the case of simpler reactions, hydrogen evolution, oxygen evolution and simple redox couples, attempts have been made to correlate the rate of the reaction with parameters such as the strength of the M—H or M—OH bond and of electron availability at the surface (Bockris and Srinivasan, 1969). Such considerations may also be used to explain the difference in products isolated from the oxidation of car-boxylic acids at platinum and carbon electrodes. At a platinum anode radical-derived products are observed (Conway and Vijh, 1967b; Eberson, 1968; Weedon, 1960) whereas at carbon the major products are derived from carbonium ion intermediates (Koehl, 1964; Sato et al., 1968). In view of recent results it must also be stated that the form of carbon is important (Brettle, 1972). Thus, the bond between the radical and platinum is strong enough to prevent its further oxidation.

In other cases, it is clear that the electrode material affects the products by the actual chemical involvement of the electrode in the chemical reactions following the electron transfer process. Thus, there can be no doubt that in reactions which produce a high current yield of an organo-metallic compound, the electrode has a dual role being both either a source or a sink for electrons and also a chemical reactant. Examples of these reactions include the reduction of alkyl bromides (Fleischmann et al., 1971a; Falli and Olivani, 1970) or the oxidation of Grignard re-agents (Braithwaite, 1966b) at lead electrodes. Both these reactions produce almost quantitative yields of the lead alkyl. Moreover, the former example is a one-electron reduction of the alkyl bromide, showing that the formation of the lead–alkyl bond is sufficiently favourable energetically to prevent further reduction, since at a more inert cathode such as mercury, a two-electron reduction occurs and only alkane is isolated as the product (Elving et al., 1955; Lantelme and Chemla, 1965; Stackelberg and Stracke, 1949).

Indeed a detailed kinetic investigation of the reduction of alkyl iodides at a lead electrode in dimethylformamide shows that the process is complex and involves catalysis by a lead alkyl species at the surface (Fleischmann et al., 1971a). The catalytic cycle proposed was

$$PbEt + EtI \xrightarrow[\text{high potential}]{\text{r.d.s.}} PbEt_2I$$

$$PbEt_2I + e \xrightarrow[\text{low potential}]{\text{r.d.s.}} PbEt_2 + I^-$$

$$PbEt_2 + EtI + e \longrightarrow PbEt_3 + I^-$$

$$2\,PbEt_3 \longrightarrow PbEt_4 + 2\,PbEt$$

where it may be seen that the catalyst is regenerated quantitatively. Confirmation of this mechanism can be obtained by carrying out preparative electrolysis on mixtures of ethyl and methyl iodides. Methyl iodide reduces at a potential 100 mV less cathodic than ethyl iodide and hence an electrolysis at a potential near the foot of the methyl iodide wave should lead to pure lead tetramethyl. In practice, however, large quantities of the ethyl radical are incorporated into the product although in the absence of methyl iodide no reduction of ethyl iodide would occur. These mixed alkyls may be formed by the catalytic mechanism proposed above because of reactions of the type

$$PbMe + EtI \longrightarrow PbEtMeI \xrightarrow{+e} I^- + PbMeEt, \text{ etc.}$$

Furthermore, it is now clear that organometallic compounds may be unstable intermediates in other electrode processes. Thus, the reduction of acetone at a series different metals in aqueous sulphuric acid has been studied (Sekine *et al.*, 1965), and the products of controlled-potential electrolyses are shown in Table 3. The reduction of isopropanol or pinacol

TABLE 3

Organic Yield (%) from the Reduction of Acetone at 1·375 V in
Aqueous Sulphuric Acid

	Hg	Pb	Zn	Cd, Al, Ni, Sn, Cu
Pinacol	2·9	7·1	0	0
Isopropanol	94·9	67·7	3·0	0
Propane	2·2	25·2	97·0	100·0

does not give propane under these conditions and, since the formation of propane is accompanied by pitting of the electrode, it was suggested that propane is formed via an unstable organometallic intermediate. The authors of this work state that this hypothesis was supported by the isolation of some stable organometallic product from the electrolyses at lead and mercury although the postulate of an organometallic intermediate may be criticized on the grounds that the yield of propane at the different metals does not reflect the expected relative ease of formation of metal–carbon bonds. Organometallic intermediates have also been postulated during the two-electron reduction of carbon–halogen bonds to alkanes at mercury electrodes. For example, a study of the reduction of 1-halo-1-methyl-2,2′-diphenylcyclopropane in acetonitrile, using cyclic voltammetry and coulometric curves, has indicated that organo-mercury compounds are intermediates (Webb *et al.*, 1970). Similar conclusions were drawn concerning the reduction of benzyl halides in dimethylformamide (Brown *et al.*, 1971; Hush and Oldham, 1963; Wawzonek *et al.*, 1964).

Although the electrode may also act as a surface for adsorption or as a reactant in a following chemical reaction, in all the examples above the electrode may basically be regarded as an "electron reagent". Early workers in the field of organic electrochemistry, however, believed that electrode reactions occurred via the production of "active hydrogen" and "active oxygen or hydroxyl radicals" followed by a chemical reduction or oxidation and, although for many years this concept fell into complete disrepute, it is now realized that, under rather special conditions, electrodes may indeed function as a source of hydrogen, hydroxyl or chlorine radicals rather than of electrons.

On anodic oxidation in aqueous alkaline solution, transition metal electrodes become covered with a monolayer or a multilayer of a passive oxide or hydroxide film, and passivated silver (Hampson *et al.*, 1969, 1970), nickel (Fleischmann *et al.*, 1971b, 1972a; Horanyi *et al.*, 1971), cobalt (Fleischmann *et al.*, 1971c) and copper (Fleischmann *et al.*, 1972a) have been investigated as electrodes for organic oxidations. At higher anodic potentials, all these oxides undergo a further oxidation to form a higher-valent oxide (in fact the transitions are $Ag^I \rightarrow Ag^{II}$, $Ni^{II} \rightarrow Ni^{III}$, $Co^{II} \rightarrow Co^{III}$ and $Cu^{II} \rightarrow Cu^{III}$), and it has been found that a wide range of organic substrates can be oxidized at these electrodes. Unusually, however, on each of the electrodes most of the substrates were oxidized at the same potential and this potential corresponded to the potential at which the oxide transition occurred. Since there was also a similarity between the electrolysis products and the products from a chemical reaction between the higher-valent oxide and the substrate, it would seem that the electrode reaction occurs via the catalytic cycle

$$M(OH)_2 \xrightarrow{-e} MO(OH) + H^+$$

$$MO(OH) + substrate \xrightarrow{r.d.s.} product + M(OH)_2$$

The currents observed for these reactions are small compared with that expected for a diffusion-controlled process and it may be shown that the current is in fact a direct measure of the rate of the chemical step in the above sequence (Fleischmann *et al.*, 1971b, 1972a). It is found that the rate of the anodic reaction decreases along the series $RNH_2 >$ $R_2NH > R_3N$, which is the opposite of that expected if electron transfer were involved in the chemical reaction. Furthermore, the products of the electrolysis and the results of kinetic experiments with deuterated methanol are more consistent with hydrogen abstraction from the carbon α to the functional group as the rate-determining step. Hence these transition metal oxide electrodes may perhaps best be considered to behave as hydroxyl radical substrates. The hydroxyl radical has also been proposed as an intermediate in the oxidation of species such as oxalic acid (Wroblowa *et al.*, 1963), ethylene (Giner, 1961; Johnson *et al.*, 1964; Shams El Din (1961) and carbon monoxide (Gilman, 1963, 1964) at low potentials on platinum. In these cases the radical is proposed solely on the basis of electrode kinetics.

Recently it was demonstrated that a platinum black-PTFE electrode, originally designed for a fuel cell, is excellent for the chlorination of double bonds and, depending on the other electrolysis conditions, it was possible to isolate the dichlorocompound or the chlorohydrin (Langer and Yurchak, 1970). Moreover, if a chlorine cathode is used, the overall process occurs with a net output of energy, i.e. the cell may do external work and the procedure has been named "electrogenerative chlorination".

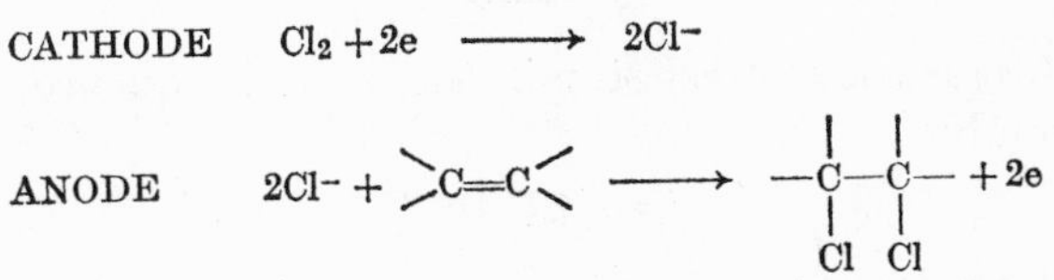

$$\text{CATHODE} \quad Cl_2 + 2e \longrightarrow 2Cl^-$$

$$\text{ANODE} \quad 2Cl^- + {>}C{=}C{<} \longrightarrow -\overset{|}{\underset{|}{C}}-\overset{|}{\underset{|}{C}}- + 2e$$
$$\qquad\qquad\qquad\qquad\qquad\qquad Cl \quad Cl$$

The anode reaction almost certainly involves adsorbed chlorine radicals.

Similar electrodes may be used for the cathodic hydrogenation of aromatic or olefinic systems (Langer and Landi, 1963, 1964), and again the cell may be used as a battery if the anode reaction is the ionization of hydrogen. Typical substrates are ethylene and benzene which certainly will not undergo direct reduction at the potentials observed at the working electrode (approximately 0·0 V versus N.H.E.) so that it must be presumed that at these catalytic electrodes the mechanism involves adsorbed hydrogen radicals.

V. Substrate Concentration

The rate of each of the steps in the overall electrode process has a simple dependence on the concentration of the electroactive species in the bulk of the solution, as in the following examples.

(i) The rate of diffusion of any species between two points is proportional to the concentration gradient between them (Fick's law). From the simplest concept of a Nernst diffusion layer, width δ, the diffusion controlled current for the reaction

$$O + ne \rightarrow R \tag{31}$$

is given by

$$i_d = -nFA(\text{flux})_O = -nFAD_O \frac{C_O^{\text{bulk}} - C_O^{\text{electrode}}}{\delta} \tag{32}$$

where D_O is the diffusion coefficient of O and A is the area of the electrode. Under the usual preparative electrolysis conditions, the reaction is carried out on the plateau of the wave on an $i\text{--}E$ curve and then $C_O^{\text{electrode}} = 0$, whence

$$i_d = -\frac{nFAD_O}{\delta} C_O^{\text{bulk}} \tag{33}$$

and the current has a first-order dependence on the concentration of the electroactive species.

(ii) When a chemical reaction (34) preceding the electron-transfer

$$\nu N \xrightarrow{\ k\ } O \tag{34}$$

step (31) is the slowest step in the overall sequence, the current is given by

$$i_k = nFAkC_N^\nu \tag{35}$$

and the order of the chemical reaction with respect to the species N is given by the slope of a $\log i_k - \log C_N$ plot.

(iii) The rate of an irreversible cathodic reduction

$$\nu O + ne \rightarrow R \tag{36}$$

is given by the equation (see Section II,A)

$$i = nFAk_1 C_O^\nu \exp\left(-\frac{\alpha nF}{RT}\phi_m\right) \tag{37}$$

Hence

$$\ln i = \ln nFAk_1 + \nu \ln C_O - \frac{\alpha nF}{RT}\phi_m \tag{38}$$

and again a plot of $\log i$ against $\log C_O$ will determine the reaction order with respect to the species O, provided that the experiment is carried out at a constant electrode potential.

(iv) When the reaction proceeds via an adsorbed reactant molecule, the current usually does not show a linear dependence on the concentration of the electroactive species. In fact a linear dependence is observed only at low coverages while Henry's law is obeyed. At high concentrations the current often becomes zero-order with respect to the substrate concentration because the surface is becoming saturated. Hence a graph of current (or coverage) versus concentration usually has the form shown in Fig. 10.

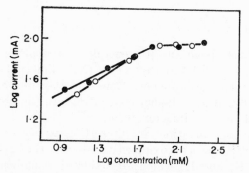

Fig. 10. $\log i$–$\log c$ plots for the oxidation of N,N-dimethylaniline in methanol at 1·20 V versus S.C.E. ● 0·5 M potassium hydroxide, ○ 0·5 M ammonium nitrate. (Taken from Weinberg and Reddy, 1968.)

It has been seen from the above simple examples that the concentration of the substrate has a profound effect on the rate of the electrode process. It must be remembered, however, that the process may show different reaction orders in the different potential regions of the i–E curve. Thus, electron transfer is commonly the slow step in the "Tafel region" and diffusion control in the plateau region and these processes may have different reaction orders. Even at one potential the reaction order may vary with the substrate concentration as, for example, in the case discussed above where the electrode reaction requires adsorption of the starting material.

Moreover, the product of an electrode process may also vary with the substrate concentration. In particular, this occurs when the reactive intermediate can further react by either a first-order or a higher-order chemical process or when the intermediate is able to react with molecules of the starting material. Examples of the latter type of reaction are certain electrochemically initiated polymerizations where clearly higher concentrations of the substrate will favour polymer of high molecular weight. The products isolated from the controlled-potential oxidation of N,N-dimethylanilines have been shown to depend on their concentration (Hand and Nelson, 1970). Thus, at low concentration the cation radical, formed as the initial product from the oxidation of N,N-dimethylaniline, undergoes simple dimerization to form mainly the p,p-coupled benzidine.

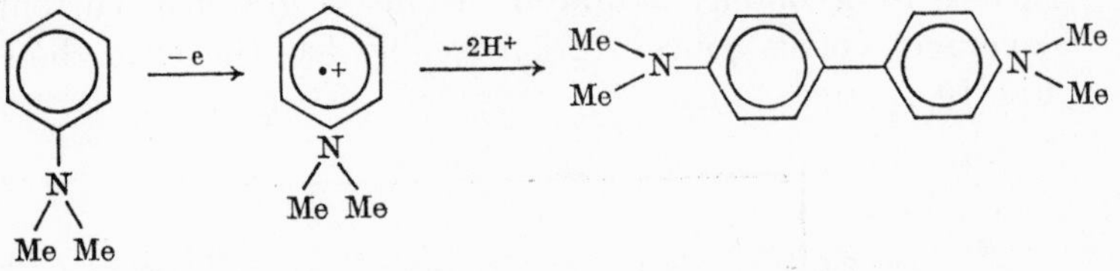

At concentrations above 10^{-3}M a complex series of reactions, ending with the formation of 4,4-methylenebis (N,N-dimethylaniline), occur. The extra methylene group comes from a further molecule of starting material. Finally, at even higher concentrations, the product becomes the dye, crystal violet or its leuco-form.

The substrate concentrations are interesting variables when a mixture of two electroactive species is oxidized or reduced. If we take the example of two species R_1 and R_2 which are oxidized at sufficiently different potentials that two clear waves are obtained on an i–E curve, an electrolysis carried out at a potential on the plateau of the first wave must occur via the route

$$R_1 - e \longrightarrow I_1$$

$$I_1 + R_2 \xrightarrow{k_1} P_1$$

whereas on the plateau of the second wave there will be competition between this route and

$$R_2 - e \longrightarrow I_2$$

$$I_2 + R_1 \xrightarrow{k_2} P_2$$

At this second potential, the ratio of the products P_1/P_2 will clearly depend on the ratio of the reactant concentrations; a large excess of R_1 will favour the former route while a large excess of R_2 will favour P_2 since the reaction between I_2 diffusing away from the electrode and incoming R_1 will ensure that the flux of R_1 at the surface is zero and hence that no I_1 will be formed. If the ratio of the reactant concentrations is of the order of unity, it can be shown that it is the ratio of the rate constants k_1/k_2 which will determine the product and since I_2, the intermediate formed at the more positive potential is invariably the least stable, it is the product P_2 which results. Thus in a heterogeneous process dominated by diffusion it is the stability of the intermediates which determines the reaction route (Fleischmann *et al.*, 1972b), whereas in homogeneous solution it is the stability of the starting materials, and this difference leads to a difference in product. These parameters have been investigated for the oxidation of a mixture of cyclohexene and chloride ion in acetonitrile (Faita *et al.*, 1969).

VI. Temperature

Although the effect of temperature on each of the steps in an overall electrode process is readily predictable, it is surprising to find in the literature very few systematic studies of this variable or attempts to use it to change the rate, products or selectivity of an organic electro-synthetic process. A recent paper has, however, discussed equipment and suitable solvents for low-temperature electrochemistry (Van Dyne and Reilley, 1972a).

The equations which describe the variation with temperature of the equilibrium constant, K_C, for a chemical system and of the rate constant, k_1, for a chemical reaction are well known. They are

$$\frac{d(\ln K_C)}{dT} = \frac{\Delta H^0}{RT^2} \tag{39}$$

and

$$\frac{d(\ln k_1)}{dT} = \frac{E_A}{RT^2} \tag{40}$$

respectively, where ΔH^0 is the enthalpy change for the forward reaction and E_A is the energy of activation. The effect of an increase of temperature on the equilibrium constant depends on whether the forward reaction is endothermic or exothermic but an increase in temperature almost always leads to a marked increase in the rate constant for the chemical reaction. Thus the above equations may be used directly to predict the effect of a temperature change on the chemical reactions preceding or following the electron transfer step. Indeed the effect of a temperature change on the limiting current for an electrode process is frequently used to distinguish whether this current is determined by the rate of diffusion or a preceding chemical reaction. Although the rate of diffusion is a function of temperature, a diffusion-limited current seldom changes by more than 1·5% per degree while a current limited by the rate of a preceding chemical reaction commonly changes by 5–10% per degree.

The use of low temperatures gives rise to the possibility of isolating reaction intermediates which, at ambient temperature, are removed by rapid chemical reactions.

Two recent papers have described the use of cyclic voltammetry at low temperatures (down to $-120°C$) to study reaction intermediates which are of very low stability at 25°C. The first (Van Dyne and Reilley, 1972b) describes the direct observation of the anion radicals of iodonitrobenzenes and the cation radicals of diphenylanthracene in dimethylformamide and of 1,2,3,6,7,8-hexahydropyrene and triphenylamine in butyronitrile at various temperatures, and the second (Byrd *et al.*, 1972) discusses the greatly increased stability of the cation radicals of aromatic hydrocarbons generated electrochemically in methylene chloride at $-70°C$. The direct calculation of activation parameters for chemical reactions following electron transfer from i–E curves has also been reported (Van Dyne and Reilly, 1972c). The use of low temperatures would introduce much additional flexibility to synthetic electrolytic techniques. For example, at very low temperatures, carbonium ions are known to be stable intermediates in strongly acidic media and an electrolytic oxidation of an alkane under these conditions could lead to a stable cation rather than the products from its reactions, as is found at room temperature. Thus the low-temperature electrolysis could lead to a variety of products, since one of a number of nucleophiles could be added at the end of the electrolysis and prior to warming to room temperature. Even a less drastic cooling of the electrolysis medium would be advantageous, since it should lead to a decrease in tarring and an increase in selectivity of the process. A decrease in temperature would also be expected to change the products from certain electrode reactions which occur by an electron transfer–chemical reaction–electron transfer (e.c.e.) mechanism at room temperature. The decrease in the rate of the chemical step would allow the one-electron product to diffuse away from the electrode before the chemical reaction occurs and the species formed by the first reaction would be the final product.

Conversely, the use of elevated temperatures will be most advantageous when the current is determined by the rate of a preceding chemical reaction or when the electron transfer occurs via an indirect route involving a rate-determining chemical process. An example of the latter is the oxidation of amines at a nickel anode where the limiting current shows marked temperature dependence (Fleischmann *et al.*, 1972a). The complete anodic oxidation of organic compounds to carbon dioxide is favoured by an increase in temperature and much fuel cell research has been carried out at temperatures up to 700°C.

The major effect of an increase in temperature on the actual electron transfer process is to increase k^0, and hence to enhance the reversibility of the electrode process. The reversible potential is, however, itself temperature dependent, and

$$\frac{dE^0}{dT} = \frac{\Delta S}{nF} \tag{41}$$

where ΔS is the entropy change per mole in the electrode reaction. Since the formation of an anion or a cation is a process accompanied by a decrease in entropy owing to the organization of the solvent around the ions and the formation of solvation shells, a decrease in temperature will favour electrode reactions where ions are formed compared with electrode reactions where uncharged species are produced from charged starting materials, i.e. if the reactions

$$A - e \rightarrow A^+$$
$$X^- - e \rightarrow X$$

occur at the same potential at room temperature the oxidation of A will be favoured compared with X^- by a decrease in temperature. For a typical entropy of ionization of -20 cal deg^{-1} mole^{-1}, a decrease in temperature of $100°C$ will separate the processes by about 100 mV. Conversely the oxidation of an anion or the reduction of a cation will be favoured by an increase in temperature. In considering the effect of temperature changes on a single electrode process it must be remembered that the potential of the reference electrode is also temperature-dependent. In this discussion we have therefore considered the variation with temperature of the difference in potential between two processes. A cell where the reference electrode is thermostatted at a single temperature and a temperature gradient exists between the reference electrode and the electrochemical cell has, however, been described (Van Dyne and Reilley, 1972a); it is claimed that the junction potential caused by this temperature gradient is only 10^{-2} mV/°C.

The exact effect of a variation of the temperature on the coverage of the electrode surface by an adsorbed species is dependent on the isotherm obeyed by the system. In general, however, a rise in the temperature of the system will be accompanied by a decrease in this coverage, and this is as would be predicted since an adsorption process would be expected to have a negative entropy.

In the case of a variable such as temperature which affects the rate of each of the steps in the overall electrode process, it is clearly necessary to have a complete understanding of the mechanism of the overall process before the total effect of a change in the parameter can be predicted. In certain cases the effect of temperature on the different steps will tend to counterbalance in the overall process. For example, an increase in temperature will increase the rate constant for a hetero- geneous chemical reaction but may also lead to a decrease in the cover- ages by the reacting species. Furthermore, although there may be clear advantages in working at low temperatures, it would still be necessary to overcome the experimental difficulties, namely low ionic solubilities, high cell resistances, low rates of diffusion and a decrease in the rate of the electron transfer.

VII. PRESSURE

While it is widely realized that pressure is a useful variable for increas- ing the solubility of the electroactive species and hence the rate of the electrode process, it is mostly forgotten that it is also a variable which affects several of the steps in the overall process. In fact these more subtle effects of pressure on organic electrode reactions do not seem to have been investigated although it is possible to estimate their import- ance by considering the known effects of pressure on chemical systems (Hamann, 1957).

The variation of the equilibrium constant K_C for the system

$$A+B \underset{k_2}{\overset{k_1}{\rightleftharpoons}} C+D$$

(when there is no change in the total number of molecules on going from reactants to products) is given by

$$\left(\frac{\partial \ln K_C}{\partial P}\right)_T = -\frac{\Delta \bar{V}^0}{RT} \tag{42}$$

where $\Delta \bar{V}^0$ is the volume change per mole which occurs during the forward reaction. The qualitative effect of pressure on the equilibrium may be predicted from the following considerations.

(i) The formation of ionic species is favoured by an increase in pressure since it is accompanied by a contraction of the solvent due to the formation of solvation shells. In fact $\Delta \bar{V}^0$ lies in the range -10 to -30 cm^3 mole^{-1} for the formation of a single complete charge.

(ii) Increased pressure also favours the formation of covalent bonds but numerically this effect is smaller than (i).

On the assumption that compressibility effects are negligible, the effect of pressure on the rate constant of the reaction

$$A+B \overset{k}{\longrightarrow} C+D$$

is given by the equation

$$\frac{\partial \ln k}{\partial P} = -\frac{\Delta V^*}{RT} \tag{43}$$

where ΔV^* is the volume change per mole when the transition state is formed from the reactants. This equation may be integrated to give

$$\log \frac{k_{P_2}}{k_{P_1}} = -\frac{\Delta V^*(P_2 - P_1)}{2 \cdot 3 RT} \tag{44}$$

but it is strictly applicable only to small pressure changes (< 100 atmospheres) since ΔV^* varies markedly with pressure. The qualitative effect of pressure on this rate constant may be predicted from the principles (i) and (ii) above, if it is possible to guess an approximate model for the transition state so that the changes in inter-atomic distances and charge separation from the reactants may be estimated.

These equations may be used directly to predict the effect of pressure on the chemical reactions preceding or following the electron transfer step and, by use of standard thermodynamic formulae, they may be modified to allow a consideration of the electron transfer step itself. For example, the electrode reaction

$$A^+ + e_{\text{electrode}} \rightarrow A$$

will show a shift in its reversible potential, ΔE^0, when the pressure is increased from 1 to P atmospheres which is given by

$$\Delta E^0 = E_P^0 - E_1^0 = -\frac{2 \cdot 3 RT}{nF} \cdot \frac{\Delta \overline{V}^0 (P - 1)}{2 \cdot 3 RT} \tag{45}$$

The formation of an ion will have a negative $\Delta \overline{V}^0$ and hence electrode reactions which produce an anion or a cation from a neutral substrate will be favoured by an increase in pressure. That is, the reversible potential for the reaction

$$B + e \rightarrow B^-$$

will become less cathodic while for the reaction

$$A \rightarrow A^+ + e$$

it will become less anodic with increasing pressure. Similarly it would seem that the rate of both these electrode processes (i.e. the rate constants for the electron transfer at the reversible potential, k^0 in equation (4)) would be enhanced by an increase in pressure since it is probable that the transition states will be more ionic in character than the substrates and hence the volumes of activation will be negative. Conversely, electrode reactions where non-ionic species are produced from ionic starting materials will become thermodynamically less favourable with increasing pressure and the rate of these electrode processes will also decrease.

It is apparent that, as in chemical systems, the magnitude of these effects will become useful and interesting from a practical viewpoint only when the pressure is increased above one kilobar. Thus for a typical electron transfer reaction with $\Delta \bar{V}^0 = -20$ cm^3 mole^{-1}, ΔE^0 will be 211 mV when the pressure is ten kilobars. This shift could be important in the not uncommon situation where, at atmospheric pressure, the oxidation of a neutral substrate occurs at around the same potential as the anion of the base electrolyte. An increase in the pressure to ten kilobars will result in a separation of the processes

$$A \rightarrow A^+ + e$$
$$X^- \rightarrow X + e$$

by almost half a volt. An analogous situation arises at the cathode when the processes

$$M^+ + e \rightarrow M$$
$$B + e \rightarrow B^-$$

occur simultaneously at atmospheric pressure. The latter process is favoured by an increase in the pressure.

An increase in pressure will also affect the rate of the diffusion of molecules to and from the electrode surface; it will cause an increase in the viscosity of the medium and hence a decrease in diffusion controlled currents. The consequences of increased pressure on the electrode double layer and for the adsorption of molecules at the electrode surface are unclear and must await investigation.

VIII. STRUCTURAL EFFECTS

A. The Rates of Simple Electrode Reactions

In Section II,A a brief account was given of the major energy terms which will determine the free energy change and thereby the electrode potential for a typical electrochemical reaction. It has also been pointed out in Section II,A that most electron transfer processes are slow so that an extra potential, the overpotential, must be applied to diminish the energy of activation in order to achieve the desired rate of reaction. It is instructive to consider the energy terms which will be important in determining the rates of electron transfer (Conway, 1965; Gurney, 1931; Horiuti and Polanyi, 1935; Levich, 1965; Marcus, 1964, 1968).

Consider first the situation for a particle approaching the surface of a metal in a vacuum (Fig. 11). If we neglect the spread of energy levels in the metal, the highest filled level with respect to the energy of the electron in vacuum will be given by ψF where ψ is the work function; similarly the lowest unfilled level in the particle will be given by the electron affinity. Let us assume next that it is possible to apply a field between the metal and vacuum as shown in Fig. 11 (for example as in

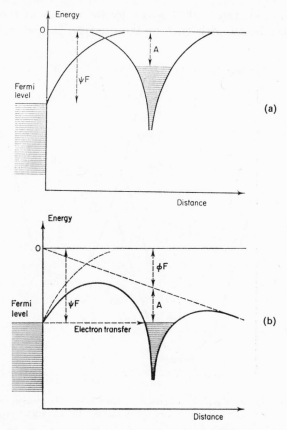

FIG. 11. The approach of an acceptor molecule to an electrode surface in a vacuum (a) in the absence of an applied field, (b) in the presence of an applied field.

field ion microscopy). Then, if the field is sufficiently high and the highest filled level in the approaching particle is depressed by ϕF so that

$$\psi F \simeq A + \phi F \qquad (46)$$

a non-radiative electron transfer will take place from the metal to the particle by a tunnel process. It should be noted that the shape and width of the barrier will depend on the applied field.

For electron transfer to and from a metal to species *in solution* the situation is complicated by the solvation energy of the species and in the general case changes in adsorption energy caused by electron transfer may also have to be taken into account. If we consider a simple reaction

$$O_{solv} + e \rightarrow R_{ads}$$

then the initial state will be given by the energy of the species O in solution combined with that of the electron in the metal; the final

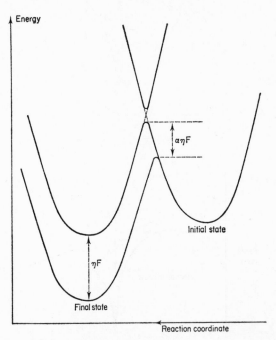

FIG. 12. Energy–reaction coordinate diagram for electron transfer in solution when there is only weak interaction between the initial and final energy states.

state, in this particular example, will be the adsorbed species R. The condition for electron transfer will then be

$$\psi F + \Delta G_O^{solv} \simeq A + \phi F + \Delta G_R^{ads} + \Delta G_R^{solv} \tag{47}$$

where we assume that R retains a residual solvation shell in the adsorbed state. Electron transfer will always take place when the energies of the initial and final states are equal. The major reaction coordinate has usually been assumed to be a change in the solvation shell. Since the electron transfer will be subject to the Franck-Condon principle, the solvation energy of the initial state must change by an appropriate amount so that the energy of the initial and final states are equal. Electron transfer will take place close to the point of intersection of the potential energy curves.

The application of the overpotential η can be considered to be equivalent to the displacement of the potential energy curves by the amount ηF with respect to each other. The high field is now applied across the double layer between the electrode and the ions at the plane of closest approach. It is apparent from Fig. 12 that the energy of activation in the favoured direction will be diminished by $\alpha\eta F$ while that in the reverse direction will be increased by $(1-\alpha)\eta F$ where the simplest interpretation of α is in terms of the slopes of the potential energy curves ($\alpha = m_1/(m_1+m_2)$) at the points of intersection: electrode processes indeed are the classical example of linear free energy relations.

In the case of redox reactions of this kind, it has been assumed that there is weak coupling between the initial and final states so that there is weak splitting (Fig. 12) and the energy of activation may be calculated from an appropriate statistical average of the states. It follows that any change in free energy of the initial state will lead to an appropriate fractional change in the energy of activation. For a given class of compounds, changes in the energy of the initial state may be estimated in the usual way, for example, from dissociation constants of substituted benzoic acids leading to the Hammett-Taft σ_R (and σ_R^*) constants for the substituents. The half-wave potential (i.e. the potential where the current reaches half the diffusion limiting value) for the reduction of electroactive compounds

$$R-\!\!\left\langle\bigcirc\right\rangle\!\!-X$$

will therefore correlate with σ_R (and σ_R^*) and give appropriate ρ_x (or ρ_x^*) values. Many examples of such structural correlations have been reported and examples are illustrated in Fig. 13 (Zuman, 1967b, 1969). The sign of the reaction constant is normally consistent with an "electrophilic reaction" with the electrode.

The observation of these linear relations is of interest since the energy changes of the initial states are obtained from a process involving displacements of atoms which are effectively equal to two-electron changes. On the other hand, electron transfer involves the movement of a single electron without atom transfer. There is no *a priori* reason therefore for the correlation. Observation of the correlations also demands that other energy terms such as the adsorption of reactants or products correlate in a linear manner with σ (or σ^*).

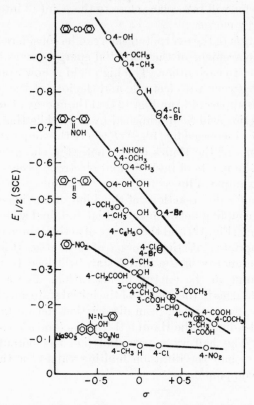

FIG. 13. Dependence of half-wave potentials for the reduction of substituted benzenes on the total polar substituent constant, σ_x. Examples shown: benzophenones [pH 0], benzophenone-oximes [pH 0], thiobenzophenones [pH 0], nitrobenzenes [pH 2·0], azo derivatives [pH 2·6]. (Taken from Zuman, 1969.)

B. *Reactions of Intermediates*

Much of the interpretation of electroorganic reactions has assumed the model implied in the above discussion, i.e. conversion of the neutral substrate into a radical ion followed by distinct chemical and/or electrochemical steps. It follows therefore that specific structural effects should be found in the reactions of the intermediates.

The most evident of these is the marked stability of radical cations formed in an aprotic medium by the oxidation of compounds where the first ionization potential (in the sense of photoelectron spectroscopy) is for the removal of an electron from a non-bonding orbital, e.g. thianthrene and diphenylene dioxide (Barry *et al.*, 1966; Cauquis and Maurey, 1968), whilst one expects and finds marked fragmentation of the molecule when the electron is removed from an orbital involved in bonding. Some stability is also found even when the electron is removed from a bonding orbital if the intermediate is able to delocalize its charge.

Secondly, the rates and modes of reaction of the intermediates are dependent on their detailed structure. For example, the stability of the cation radical formed by the oxidation of tertiary aromatic amines is markedly dependent on the type and degree of substitution in the *p*-position (Adams, 1969b; Nelson and Adams, 1968; Seo *et al.*, 1966), and the rate of loss of halogen from the anion radical formed during the reduction of haloalkyl-nitrobenzenes is dependent on the size and position of alkyl substituent and the increase in the rate of this reaction may be correlated with the degree to which the nitro group is twisted out of the plane of the benzene ring (Danen *et al.*, 1969).

Thirdly, most of the structural effects familiar from reactions of charged intermediates in homogeneous solution have been observed in electrode reactions. For example, typical carbonium ion rearrangements, the neopentyl rearrangement (Miller and Hoffman, 1967), hydride shifts (Fleischmann *et al.*, 1970), ring contractions (Baggaley and Brettle, 1968), ring expansions (Corey *et al.*, 1960) and ring closures (Koehl, 1964) all occur. It would also be expected that it should be possible to form intermediates which undergo characteristic electro-cyclic reactions. Such reactions have been proposed and found in homogeneous reactions (Woodward and Hoffmann, 1968).

Second-order effects of structure on reactions succeeding electron transfer would also be expected. As the intermediates are formed in a region having a very high field, changes in the charge distribution of radical ions should take place with changes of the potential and these should affect the sites of the succeeding steps (such as protonation) and lead to a consequent alteration in product distribution (Hoijtink, 1957).

C. *Reaction Coordinates*

It has already been noted that the changes in the solvent structure of the initial and final states in the reaction are the major factors determining the energy of activation. The magnitudes of these changes in the solvation energy should in turn be dependent on the structure of the organic molecules, in the first place due to the charge delocalization in the final state. Some support for this view is provided by the fact that the rates of formation of radical ions from polynuclear aromatic hydrocarbons (where the charge will be strongly delocalized and the solvent loosely bound) are high (Hale, 1971; Peover, 1971). Unfortunately the rates of most of these species are indeed so large that they are at the limit of current methods of measurement and systematic work has been restricted (Hoijtink and Aten, 1961; Peover and Dietz, 1968). Charge localization by suitable substitution, such as for nitro compounds (Peover and Powell, 1969), on the other hand, leads to a marked lowering in the rate of reaction and the magnitude of the change is in agreement with predictions.

In the case of many classes of organic molecules it would also be expected that major internal changes in the molecules would be associated with the electron transfer, and such structural changes could well become the major coordinates determining the rates. At present there is

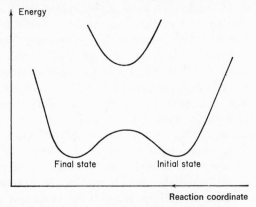

FIG. 14. Energy–reaction coordinate diagram for electron transfer in solution when there is strong interaction between the initial and final energy states.

only little information with regard to this view. Some support is provided by the decrease in rate of reduction of a series of stilbenes with increasing steric hindrance caused by change of isomer or by increasing substitution (Peover and Dietz, 1968). It is likely that changes in the configuration of the non-planar parent molecule must take place in forming the radical anion and these will be increasingly retarded with increase of the degree of steric hindrance.

For the mechanisms considered so far it has been assumed that electron transfer precedes any change in composition of the molecule. It is possible, however, that changes in composition are concerted with the electron transfer and in these cases of combined atom and electron transfer it will not be appropriate to estimate the energy of activation by carrying out an appropriate statistical average of the initial and final states in the manner of Fig. 12. A similar situation would arise in the case of electron transfer alone if there is strong coupling between the initial and final states (see Fig. 14) with a consequent large degree of splitting.

There have been a number of investigations of the formulation of the problem of electron transfer accompanied by atom transfer particularly with regard to the simultaneous movement of the proton (which, in view of its small mass, may in fact be an atypical case). A possible model for such processes would assume a conservation of bond order along the reaction coordinates (Johnston, 1960). It is of interest that the results of such calculations are similar to those for electron transfer for weak coupling, although the interpretation of the process and parameters (such as α) are different.

Concerted movement of atoms and electrons might be expected to be of several kinds. There could be a bimolecular rate-determining reduction $B + H^+ \xrightarrow{e} BH\cdot$, or the converse $RH \xrightarrow{-e} R\cdot + H^+$ (rather than $RH \xrightarrow{-e} RH^+$), or the detachment of other "leaving groups" or even more complex molecular motion

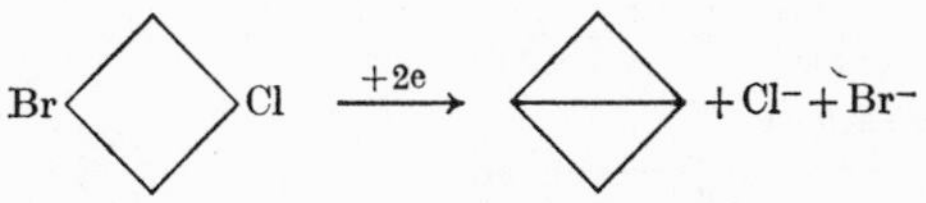

where reduction takes place at the potential appropriate to the weakest carbon halogen bond, i.e., C—Br (Rifi, 1969).

A difficulty in the analysis of reactions suspected to take place with concerted movement of atoms lies in the irreversibility of most of the processes and the consequent impossibility of characterizing rates at the reversible potential. An approach which has been suggested is the comparison of the substituent effects of appropriate electrode reductions and homogeneous reactions as in the example (Marcus, 1968)

$$XC_6H_4.CO.CH_3 + H^+ + e \rightarrow XC_6H_4\dot{C}(OH)CH_3$$

with

$$XC_6H_4.CO.CH_3 + R\cdot \rightarrow XC_6H_4.\dot{C}(OR)CH_3$$

IX. Cell Design

In the laboratory, preparative electrolyses on the one gram scale can readily be carried out in simple three-electrode cells. The connection of such a cell to a typical potentiostat (feedback system) is illustrated in Fig. 15. It is normally desirable that the electrolysis should be carried out at constant temperature and potential and at a high rate. Hence when designing such cells it is necessary to consider a number of factors. These include the following.

(i) The working and auxiliary electrodes should be placed as near as possible so as to minimize the cell resistance. This maximizes the current which may be driven through the cell with a particular potentiostat and reduces the resistive heating of the cell.

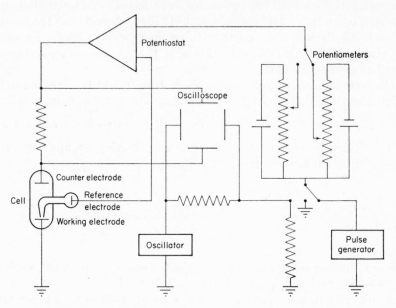

FIG. 15. Simple electronic circuit for a controlled potential experiment.

(ii) The working and auxiliary electrodes should be placed in a configuration such that all points on the surface of the working electrode are similar with respect to the auxiliary electrode so that there is a uniform distribution of the current density; only under these circumstances will the whole electrode surface be at the same potential. Therefore, for example, the use of a large plate working electrode with only one side facing the auxiliary electrode should be avoided since the back surface of the electrode will be at a different potential from the front face and, indeed, a different electrode reaction may occur.

(iii) The Luggin capillary probe leading to the reference electrode should be placed close to the working electrode (< 1 mm) and in the main current path so that the potential is probed at a typical point on the surface.

(iv) The surface area of the working electrode (or more exactly the ratio of the area of the electrode to the volume of the electrode compartment) should be large so that the electrolysis is carried out rapidly. Additionally, this reduces the importance of slow chemical reactions which might adversely affect the current yield of the desired product.

(v) The contact time between the electrode and the reactant can be important in determining the product, particularly when an e.c.e. mechanism occurs in an unstirred solution. An example, given earlier in this chapter, is the oxidation of 9,10-diphenylanthracene in the presence of 4-methylpyridine which yields a one-electron product in highly stirred solution but a two-electron product in unstirred conditions (Manning *et al.*, 1969). In the case of diffusion-limited currents, the electrolysis time is also decreased by stirring. In these small preparative cells, a high rate of stirring is most easily obtained by use of an inert gas, although it is also possible to arrange for the working electrode to rotate or vibrate.

(vi) Electrolyses are commonly carried out with a separator between the working and auxiliary electrodes. This separator is designed to prevent mixing or reaction between the products formed at the working and auxiliary electrodes and to prevent the product formed at the working electrode being removed at the auxiliary electrode. The commonest cause of a change in the product when an electrolysis is carried out in an undivided instead of a separated cell is the pH change induced at the auxiliary electrode. Anodic oxidations frequently produce protons whilst conversely cathodic reactions can remove protons or produce a basic species. The barrier prevents the free exchange of these species between the anode and cathode compartment. Examples include the indirect oxidation of olefins in aqueous chloride (Ibl and Selvig, 1970; Kellogg Co., 1963; Le Duc, 1968) and the reduction of benzene in lithium chloride/methylamine (Benkeser and Kaiser, 1963; Benkeser *et al.*, 1964). In the former example the product is the chlorohydrin in a separated cell

$$Cl^- + H_2O \rightarrow HClO + H^+ + 2e$$
$$RCH=CH_2 + HClO \rightarrow \underset{\underset{OH}{|}}{RCH}-\underset{\underset{Cl}{|}}{CH_2}$$

but the olefin oxide in an unseparated cell, since hydroxide ion produced at the cathode

$$H_2O + e \rightarrow \tfrac{1}{2}H_2 + OH^-$$

is able to react with chlorohydrin to produce the oxide. In the latter example the product in the undivided cell is the unconjugated diene but cyclohexene in a separated cell. This difference has been explained by the observation that lithium methylamide will catalyze the conversion of the unconjugate dienes to its conjugated isomer. Therefore the cyclohexene is formed by the route

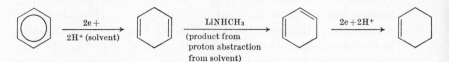

whereas in an undivided cell protons are produced in the anode reaction and therefore no lithium methylamide is formed and the reaction stops at the first step.

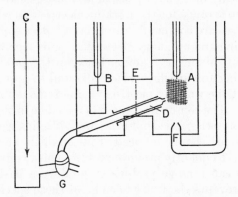

FIG. 16. Small-scale laboratory cell for preparative electrolysis. A, Pt gauze working electrode. B, Pt sheet secondary electrode. C, Reference electrode. D, Luggin capillary on a syringe barrel so that the position of the tip of the Luggin probe relative to the working electrode is readily adjustable. E, Glass sinter to separate anode and cathode compartments. F, Gas inlet to allow stirring with inert gas or the continuous introduction of reactant. G, Three-way tap where a boundary between the reference electrode and the working solutions may be formed.

The separator is frequently a sintered glass frit, but it may also be any of a wide range of inert, porous materials such as celloton, vycor or porvic or an ion exchange membrane. A number of stable ion exchange membranes suitable for use in aqueous and non-aqueous solvents have become available in recent years.

Figure 16 shows a diagram of a small cell, working electrode compartment 10 cm³, which combines many of the features discussed above.

For an electrosynthetic process on a much increased scale in the laboratory or on a commercial industrial scale, the cell design is much more critical and it will be markedly different from that used for small-scale experiments in the laboratory. The design features of an ideal industrial cell have been listed as follows (Goodridge, 1968).

(a) Low capital cost.
(b) Uniform electrode potential.
(c) Large ratio of electrode surface area to solution volume.
(d) Low cell voltage (i.e., low cell resistance).
(e) Good heat and mass transfer.
(f) Simplicity of construction and ease of replacement of cell parts.
(g) Capability for handling gaseous reactants or products.
(h) Suitability for use at elevated pressures.
(i) Ease of continuous operation.

For an industrial application it is clear that flow cells will be highly advantageous since they allow continuous operation, the use of external heat exchangers and continuous monitoring of quality. In addition the reactant concentration remains invariant with time, allowing the cell to be run under true steady-state conditions, and the lower residence time of the products reduces further reactions.

Earlier designs consisted of a stack of plane electrode cells arranged in the form of a conventional filter press. The cells are connected in series and the electrolyte flow is usually at right angles to the current path. The electrodes are insulated from each other by gaskets and the anolyte and catholyte streams may be separated by incorporating a diaphragm. This type of cell has several advantages but it suffers from being expensive to construct and to replace parts, difficult to thermostat effectively and unsuitable for obtaining very high rates of mass transfer. Filter press cells have, however, been used in recent years for the Monsanto adiponitrile process (Baizer, 1964).

More recently, a number of less conventional cells based on the principle of using very thin gaps between the anode and the cathode have been described (Beck and Guthke, 1969) and these have the advantages that they may be used in poorly conducting media and that very high rates of mass transfer may be induced. A number of these configurations consist simply of two electrodes placed in very close proximity [0·01 to 1 mm] and either the electrolyte is pumped through the gap or one of the electrodes is vibrated in a direction perpendicular to the plane of the electrodes with a frequency of 5–100 Hz and an amplitude of 0·02 to 2·0 mm. A bipolar capillary cell has also been described (Beck and Guthke, 1969). It consists of a stack of carbon plates separated from one another by radially arranged 125 μm polyester foils. The electrolysis solution flows through the capillaries and a d.c.

potential is applied across the end plates. The cell shown diagrammatically in Fig. 17 has been reported to be suitable for the hydrodimerization of acrylonitrile, and adiponitrile was obtained in 90% yield.

A number of recent cell designs have used particulate electrodes. The early versions had slurry electrodes (Butler and Fawcett, 1964; Gerischer, 1963) but more recently the electrodes have been packed or

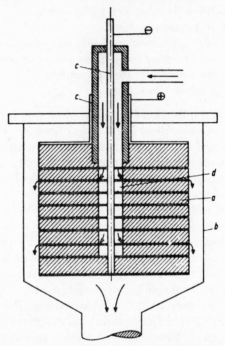

FIG. 17. Bipolar capillary cell: (a) 125 μm gaps, (b) glass vessel, (c) electrical leads, (d) stack of graphite plates, (e) solution flow. (Taken from Beck and Guthke, 1969.)

fluidized beds. The Nalco process for the production of lead alkyls from the oxidation of Grignard reagents in a mixture of ethers uses an anode which is a packed bed of lead pellets (Braithwaite, 1966a, b). The anode is contained in a tubular mesh which is surrounded by a diaphragm and by the cathode. A recent paper has reported a study of the potential distribution in a packed bed electrode (Goodridge and Ismail, 1971).

Fluid bed electrodes consist of a bed of particles supported by a structure such as a coarse sinter and fluidized by an upward stream of electrolyte and two different configurations have been described where the current path is parallel or perpendicular to the direction of fluidization (Backhurst et al., 1969). Such electrodes have been used for electrosynthetic reactions and, in particular, a pilot plant for the reduction of *m*-nitrobenzenesulphonic acid to metanilic acid has been run. A cathode of copper particles was used in a 1000 A unit and the overall performance of the electrode was satisfactory (Smith, 1969).

During the last year a bipolar particular cell has been described (Fleischmann *et al.*, 1971d). The cell shown in Fig. 18 consists of a packed bed of mixed conducting and non-conducting particles; a d.c. potential is applied between two feeder electrodes situated at the ends

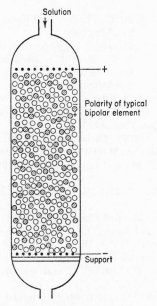

FIG. 18. Bipolar particulate cell. ● conducting particles, ○ non-conducting glass beads.

of the bed and, provided a dilute electrolyte solution [$< 10^{-2}$ M] is used, the difference in conductivity between the electrolyte and the conducting particles ensures that each conducting particle becomes a bipole. With a bed of ballotini glass spheres (diameter 0·5 mm), with a small proportion covered with a conducting graphite, it has been shown that propylene may be oxidized to propylene oxide using the bromide/hypobromite couple as catalyst. The throughput per unit volume is comparable to conventional cells working with concentrated electrolyte. Furthermore, the cell uses power at high voltage and low current and its construction lends itself to an assembly of parallel tubular units. Such reactors would be very easy to construct and should have low capital costs. The possibility of carrying out syntheses using low concentrations of electrolytes also promises a simplification of work up processes.

REFERENCES

Adams, R. N. (1969a). "Electrochemistry at Solid Electrodes", Marcel Dekker, New York.
Adams, R. N. (1969b). *Acc. Chem. Res.* **2**, 175.
Anderson, J. D., Petrovich, J. P., and Baizer, M. (1969). *Adv. Org. Chem.* **6**, 257.
Arapakos, P. G., and Scott, M. K. (1968). *Tetrahedron Lett.* 1975.
Avaca, A., and Bewick, A. (1971). "Proceedings of Electrochemical Engineering Symposium", Newcastle. To be published by the Institution of Chemical Engineers (1973).
Backhurst, J. R., Coulson, J. M., Goodridge, F., Plimley, R. E., and Fleischmann, M. (1969). *J. Electrochem. Soc.* **116**, 1600.
Baggaley, A. J., and Brettle, R. (1968). *J. Chem. Soc.* (C), 2055.
Bagotzky, V. S., and Vassilyev, Yu. B. (1967). *Electrochim. Acta* **12**, 1323.
Baizer, M. (1964). *J. Electrochem. Soc.* **111**, 215.
Baizer, M. (1969). *Naturwissenschaften* **56**, 405.
Baizer, M., and Petrovich, J. P. (1970). *Prog. Phys. Org. Chem.* **7**, 189.
Barry, C., Cauquis, G., and Maurey, M. (1966). *Bull. Soc. Chim. France* 2510.
Beck, F. (1970). *Chem. Ing. Tech.* **42**, 153.
Beck, F., and Guthke, H. (1969). *Chem. Ing. Tech.* **41**, 943.
Benkeser, R. A., and Kaiser, E. M. (1963). *J. Am. Chem. Soc.* **85**, 2258.
Benkeser, R. A., Kaiser, E. M., and Lambert, R. F. (1964). *J. Am. Chem. Soc.* **86**, 5272.
Benkeser, R. A., Watanabe, H., Mels, S. J., and Sabol, M. A. (1970). *J. Org. Chem.* **35**, 1210.
Bertram, J., Fleischmann, M., and Pletcher, D. (1971). *Tetrahedron Lett.* 369.
Bessard, J., Cauquis, G., and Serve, D. (1970). *Tetrahedron Lett.* 3103.
Bewick, A., and Pletcher, D. (1970, 1971). "Specialist Periodical Reports of the Chemical Society—Electrochemistry" (G. J. Hills, ed.).
Bhattacharyya, D. N., Lee, C. C., Smid, J., and Szwarc, M. (1965). *J. Phys. Chem.* **69**, 608.
Blomgren, E., Bockris, J. O'M., and Jesch, C. (1961). *Chem.* **65**, 2000.
Bobbitt, J. M. (1972). Personal communication.
Bobbitt, J. M., Noguchi, I., Jagi, H., and Weisgraber, K. H. (1971). *J. Am. Chem. Soc.* **93**, 3551.
Bockris, J. O'M. (1954). "Modern Aspects of Electrochemistry", Butterworths, **1**, 180.
Bockris, J. O'M., Devanathan, M. A. V., and Müller, K. (1963). *Proc. Royal Soc.* **274A**, 55.
Bockris, J. O'M., and Srinivasan, S. K. (1969). "Fuel Cells—Their Electrochemistry", McGraw Hill.
Braithwaite, D. (1966a). U.S. Patent 3,256,161.
Braithwaite, D. (1966b). U.S. Patent 3,312,605.
Brettle, R. (1972). Euchem Meeting—Ronneby, Sweden.
Brown, E. R., and Large, R. F. (1971). "Physical Methods of Chemistry—Part IIa" (A. Weissberger and B. W. Rossiter, eds.), John Wiley & Sons, 523.
Brown, O. R., and Harrison, J. A. (1969). *J. Electroanal. Chem.* **21**, 387.
Brown, O. R., Thirsk, H. R., and Thornton, B. (1971). *Electrochim. Acta* **16**, 495.
Buschow, K. H. J., Dieleman, J., and Hoijtink, G. J. (1965). *J. Chem. Phys.* **42**, 1993.
Butler, R. M., and Fawcett, W. R. (1964). *Canad. Patent* 200, 933.

Byrd, L., Miller, L. L., and Pletcher, D. (1972). *Tetrahedron Lett.* 2419.

Cauquis, G. (1966). *Bull. Soc. Chim. France* 457.

Cauquis, G., and Genies, M. (1970). *Tetrahedron Lett.* 3403.

Cauquis, G., and Maurey, M. (1968). *Compt. Rend. C.* **266**, 1021.

Chang, J., Large, R. F., and Popp, G. (1971). "Physical Methods of Chemistry—Part IIb" (A. Weissberger and B. W. Rossiter, eds.), John Wiley and Sons, 1.

Christie, J. H. (1967). *J. Electroanal. Chem.* **13**, 79, 227.

Clark, D., Fleischmann, M., and Pletcher, D. (1972). *J. Electroanal. Chem.* in press.

Coetzee, J. F., McGuire, D. K., and Hedrick, J. L. (1963). *J. Phys. Chem.* **67**, 1814.

Coleman, J., Fleischmann, M., and Pletcher, D. (1973). *J.C.S. Perkin II*, 374.

Conway, B. E. (1965). "Theory and Principles of Electrode Processes", Ronald Press Co.

Conway, B. E., and Vijh, A. K. (1967a). *Z. Analyt. Chem.* **224**, 149, 160.

Conway, B. E., and Vijh, A. K. (1967b). *Chem. Reviews* **67**, 623.

Conway, B. E., Rudd, E. J., and Gordon, L. G. M. (1968). *Disc. Faraday Soc.* **45**, 87.

Corey, E. J., Bauld, N. L., La Londe, R. T., Casanova, J., and Kaiser, E. T. (1960). *J. Am. Chem. Soc.* **82**, 2645.

Damaskin, B. B. (1967). "The Principles of Current Methods for the Study of Electrochemical Reactions", McGraw-Hill.

Danen, W. C., Kensler, T. T., Lawless, J. G., Marcus, M. F., and Hawley, M. D. (1969). *J. Phys. Chem.* **73**, 4389.

Delahay, P. (1961). "Advances in Electrochemistry and Electrochemical Engineering", Interscience, **1**, 233.

Delahay, P. (1966). "Double Layer and Electrode Kinetics", Interscience, 256.

de Nova, V. (1970). "Proceedings of Electrochemical Society Spring Meeting", Los Angeles.

Doughty, A., Fleischmann, M., and Pletcher, D. (1972). Unpublished results.

Dubois, J. E., and Dodin, G. (1969). *Tetrahedron Lett.* 2325.

Dutkiewicz, E., and Parsons, R. (1966). *J. Electroanal. Chem.* **11**, 196.

Eberson, L. (1968). "Chemistry of the Carboxyl Group", (S. Patai, ed.), Interscience, New York.

Eberson, L., and Olafson, B. (1969). *Acta Chem. Scand.* **23**, 2355.

Eberson, L., and Parker, V. D. (1969). *Tetrahedron Lett.* 2839, 2843.

Eberson, L., and Schäffer, H. (1971). "Organic Electrochemistry", Springer-Verlag Press.

Elving, P. J., and Hilton, C. L. (1952). *J. Am. Chem. Soc.* **74**, 3368.

Elving, P. J., Rosenthal, I., and Martin, A. J. (1955). *J. Am. Chem. Soc.* **77**, 5218.

Faita, G., Fleischmann, M., and Pletcher, D. (1969). *J. Electroanal. Chem.* **25**, 455.

Feokstistov, L. G. (1968). "Progress in Electrochemistry of Organic Compounds" (A. N. Frumkin and A. B. Ershler, eds.), Plenum Press, **1**, 135.

Fichter, F. (1942). "Organische Elektrochimie", Steinkopf, Dresden.

Fioshin, M. Ya., and Avrutskaya, I. A. (1967). *Electrokhimiya* **3**, 1288.

Fleischmann, M., and Pletcher, D. (1968). *Tetrahedron Lett.* 6255.

Fleischmann, M., and Pletcher, D. (1969). *R.I.C. Reviews* **2**, 87.

Fleischmann, M., and Pletcher, D. (1971). "Reactions of Molecules at Electrodes" (N. Hush, ed.), John Wiley and Sons, 347.

Fleischmann, M., and Pletcher, D. (1972). Unpublished work.

Fleischmann, M., Petrov, I. N., and Wynne-Jones, W. F. K. (1964). "Electro-chemistry—Proceedings of 1st Australian Conference", Pergamon, 500.

Fleischmann, M., Mansfield, J. R., and Lord Wynne-Jones (1965). *J. Electroanal. Chem.* **10**, 522.

Fleischmann, M., Pletcher, D., and Race, G. M. (1969). *J. Electroanal. Chem.* **23**, 369.

Fleischmann, M., Pletcher, D., and Race, G. M. (1970). *J. Chem. Soc.* (B), 1746.

Fleischmann, M., Pletcher, D., and Vance, C. J. (1971a). *J. Electroanal. Chem.* **29**, 325.

Fleischmann, M., Korinek, K., and Pletcher, D. (1971b). *J. Electroanal. Chem.* **31**, 39.

Fleischmann, M., Korinek, K., and Pletcher, D. (1971c). *J. Electroanal. Chem.* **33**, 478.

Fleischmann, M., Oldfield, J. and Tennakoon, C. T. K. (1971d). "Proceedings of Electrochemical Engineering Symposium", Newcastle. To be published by the Institution of Chemical Engineers (1973).

Fleischmann, M., Korinek, K., and Pletcher, D. (1972a). *J.C.S. Perkin II*, 1396.

Fleischmann, M., Joslin, T., and Pletcher, D. (1972b). Unpublished results.

Fleischmann, M., Pletcher, D., and Vance, C. J. (1972c). *J. Organometallic Chemistry* **40**, 1.

Fleischmann, M., Korinek, K., and Pletcher, D. (1972d). *J. Electroanal. Chem.* **34**, 499.

Franklin, T. C., and Sothern, R. D. (1954). *J. Phys. Chem.* **58**, 951.

Frumkin, A. N. (1966). *J. Chim. Phys.* **63**, 785.

Frumkin, A. N. (1961) *Electrochim. Acta*, **5**, 265.

Fry, A. J., Mitnick, M., and Reed, R. G. (1970). *J. Org. Chem.* **35**, 1232.

Galli, R., and Olivani, F. (1970). *J. Electroanal. Chem.* **25**, 331.

Gerischer, H. (1963). *Ber. Bunsen Ges. Phys. Chem.* **67**, 164.

Gerischer, H., and Vielstich, W. (1955). *Z. Phys. Chem.* **3**, 16; **4**, 10.

Gillet, I. E. (1961). *J. Electrochem. Soc.* **108**, 71.

Gillett, I. E. (1968). *Chem. Ing. Techn.* **40**, 573.

Gilman, S. (1963). *J. Phys. Chem.* **67**, 78, 1898.

Gilman, S. (1964). *J. Phys. Chem.* **68**, 70.

Gilman, S. (1967). *Electroanal. Chem.* **2**, 111.

Giner, J. (1961). *Electrochim. Acta* **4**, 42.

Glasstone, S., and Hickling, A. (1934). *J. Chem. Soc.* 1878.

Goodridge, F. (1968). *Chem. Proc. Eng.* **49**, 93.

Goodridge, F., and Ismail, B. M. (1971). "Proceedings of Electrochemical Engineering Symposium", Newcastle.

Grabowski, Z. R., Czochralska, B., Vincenz-Chodkowska, H., and Balasiewicz, M. S. (1968). *Disc. Faraday Soc.* **45**, 145.

Gurney, R. W. (1931). *Proc. R. Soc. Lond.* **A134**, 137.

Haber, F. (1898). *Z. Elektrochem.* **4**, 506.

Hale, J. (1971). "Reactions of Molecules at Electrodes" (N. S. Hush, ed.), 229.

Hamann, S. D. (1957). "Physico-Chemical Effects of Pressure", Butterworths.

Hampson, N. A., Lee, J. B., Morley, J. R., and Scanlon, B. (1969). *Can. J. Chem.* **47**, 3729.

Hampson, N. A., Lee, J. B., Morley, J. R., MacDonald, K. I., and Scanlon, B. (1970) *Tetrahedron* **26**, 1109.

Hand, R., and Nelson, R. F. (1970). *J. Electrochem. Soc.* **117**, 1353.

Headridge, J. B., and Pletcher, D. (1967). *Inorg. Nucl. Chem. Lett.* **3**, 475.

Hoijtink, G. J. (1957). *Rec. Trav. Chim.* **76**, 869, 885.
Hoijtink, G. J., and Aten, A. C. (1961). "Advances in Polarography", Oxford, 771.
Horányi, G., Vertés, G., and Nagy, F. (1971). *Acta Chim. Hung.* **67**, 357.
Horiuti, J., and Polanyi, M. (1935). *Acta Phys.-chim. URSS* **2**, 505.
Horner, L. (1970). *Chem. Ing. Techn.* **42**, 189.
Horner, L., and Degner, D. (1971). *Tetrahedron Lett.* 1241.
Horner, L., and Skaletz, D. H. (1971). *Tetrahedron Lett.* 3679.
Hush, N. S., and Oldham, K. B. (1963). *J. Electroanal. Chem.* **6**, 34.
Ibl, N., and Selvig, A. (1970). *Chem. Ing. Techn.* **42**, 180.
Johnson, J. W., Wroblowa, H., and Bockris, J. O'M. (1964). *Electrochim. Acta* **9**, 639.
Johnston, H. S. (1960). *Adv. Chem. Phys.* **3**, 131.
Kanakam, P., Pathy, M. S. V., Udupa, H. W. V. (1967). *Electrochim. Acta* **13**, 329.
Kargin, J. H., Manoušek, O., and Zuman, P. (1966). *J. Electroanal. Chem.* **12**, 443.
Kellogg Co. (1963). Belgian Patent 637691.
Koch, D. F. A., and Woods, R. (1968). *Electrochim. Acta* **13**, 2101.
Koehl, W. J. (1964). *J. Am. Chem. Soc.* **86**, 4684.
Kollmar, H., and Smith, H. O. (1970). *Chem. Phys. Lett.* **5**, 7.
Kolthoff, I. M., and Coetzee, J. F. (1957). *J. Am. Chem. Soc.* **79**, 1852.
Korchinskii, G. A. (1968). *Ukrainsk Zhim. Zh.* **29**, 103.
Koryta, J., Dvořak, J., and Boháčková, V. (1966). "Electrochemistry", Methuen and Co.
Kuwana, T., and French, W. G. (1964). *Anal. Chem.* **36**, 241.
Langer, S. H., and Landi, H. P. (1963). *J. Am. Chem. Soc.* **85**, 3043.
Langer, S. H., and Landi, H. P. (1964). *J. Am. Chem. Soc.* **86**, 4694.
Langer, S. H., and Yurchak, S. (1970). *J. Electrochem. Soc.* **117**, 510.
Lantelme, F., and Chemla, M. (1965). *Electrochim. Acta* **10**, 657.
Le Duc, J. A. M. (1968). U.S. Patent 337,627.
Levich, V. G. (1965). "Advances in Electrochemistry and Electrochemical Engineering", **4**, 249.
Lund, H. (1966). "Proceedings of 4th International Congress on Polarography", Prague.
Lund, H. (1967). *Österr. Chem. Zeitung* **68**, 43, 152.
Mann, C. K. (1969). *Electroanal. Chem.* **3**, 58.
Mann, C. K., and Barnes, K. K. (1970). "Electrochemical Reactions in Non-Aqueous Systems", Marcel Dekker.
Manning, G., Parker, V. D., and Adams, R. N. (1969). *J. Am. Chem. Soc.* **91**, 4584.
Marcus, R. A. (1964). *Ann. Rev. Phys. Chem.* **15**, 155.
Marcus, R. A. (1968). *Disc. Faraday Soc.* **45**, 7.
Masui, M., Sayo, H., and Tsuda, Y. (1968). *J. Chem. Soc. (B)*, 973.
Mehl, W., and Hale, J. M. (1968). *Disc. Faraday Soc.* **45**, 30.
Meites, L. (1965). "Polarographic Techniques", Interscience.
Meites, L. (1971). "Physical Methods of Chemistry—Part IIa" (A. Weissberger and B. W. Rossiter, eds.), John Wiley and Sons, 646.
Miller, L. L., and Hoffman, A. K. (1967). *J. Am. Chem. Soc.* **89**, 593.
Miller, L. L., and Riekena, E. (1969). *J. Org. Chem.* **34**, 3359.
Miller, L. L., and Mayeda, E. A. (1970). *J. Am. Chem. Soc.* **92**, 5819.
Miller, L. L., and Mayeda, E. A. (1972). Private communication.
Miller, L. L., Koch, V. R., Larshcheid, M. E., and Wolf, J. F. (1971). *Tetrahedron Lett.* 1389.
Nelson, R. F., and Adams, R. N. (1968). *J. Am. Chem. Soc.* **90**, 3925.

Nicholson, R. S., and Shain, I. (1966). *Anal. Chem.* **36**, 706.

Nürnberg, H. W. (1965). *Disc. Faraday Soc.* **39**, 136.

O'Connor, J. J., and Pearl, I. A. (1964). *J. Electrochem. Soc.* **111**, 335.

Olah, G. A., and Pittman, C. U. (1966). *Adv. Phys. Org. Chem.* **4**, 303.

Olah, G. A., Bollinger, J. M., Cupus, C. A., and Lukas, J. (1967). *J. Am. Chem. Soc.* **89**, 2692.

Olah, G. A., Halpern, Y., Shen, J., and Mo. Y. K. (1971). *J. Am. Chem. Soc.* **93**, 1251.

Parker, V. D. (1969). *Chem. Commun.* 848.

Parker, V. D. (1970). *Acta Chem. Scand.* **24**, 3455.

Parsons, R. (1958). *Trans. Faraday Soc.* **54**, 1053.

Parsons, R. (1968). *Disc. Faraday Soc.* **43**, 40.

Payne, R. (1967). *J. Phys. Chem.* **71**, 1548.

Payne, R. (1970). "Advances in Electrochemistry and Electrochemical Engineering", **7**, 1.

Peover, M. (1964). *Trans. Faraday Soc.* **60**, 417.

Peover, M. (1967). *Electroanal. Chem.* **2**, 1.

Peover, M. (1971). "Reactions of Molecules at Electrodes" (N. S. Hush, ed.), John Wiley and Sons, 259.

Peover, M. E., and Davies, J. D. (1963). *J. Electroanal. Chem.* **6**, 46.

Peover, M. E., and Davies, J. D. (1964). "Polarography" (Hills, ed.), Macmillan, 1003.

Peover, M. E., and Dietz, R. (1968). *Disc. Faraday Soc.* **45**, 154.

Peover, M., and Powell, J. S. (1969). *J. Electroanal. Chem.* **20**, 427.

Peover, M., Allison, A. C., and Gough, T. A. (1963). *Nature* **197**, 764.

Peover, M., Case, B., Hush, N. S., and Parsons, R. (1965). *J. Electroanal. Chem.* **10**, 360.

Petrii, O. A. (1969). "Progress in Electrochemistry of Organic Compounds" (A. N. Frumkin and A. B. Ershler, eds.), Plenum Press, **1**, 319.

Piekarski, S., and Adams, R. N. (1971). "Physical Methods of Chemistry—Part IIa" (A. Weissberger and B. W. Rossiter, eds.), John Wiley & Sons, 531.

Pletcher, D. (1967). Ph.D. Thesis. Univ. of Sheffield.

Reeves, R. M. (1969). Ph.D. Thesis, Univ. of Southampton.

Riddiford, A. C. (1968). "Advances in Electrochemistry and Electrochemical Engineering" **4**, 47.

Rifi, M. R. (1969). *Tetrahedron Lett.* 1043.

Rifi, M. R. (1971). *J. Org. Chem.* **36**, 2017.

Sato, N., Sekine, T., and Sugino, T. (1968). *J. Electrochem. Soc.* **115**, 242.

Schwarz, W., and Shain, I. (1965). *J. Phys. Chem.* **69**, 30.

Sekine, T., Yamura, A., and Sugino, T. (1965). *J. Electrochem. Soc.* **112**, 439.

Seo, T. E., Nelson, R. F., Fritsch, J. M., Marcoux, L. S., Leedy, D. W., and Adams, R. N. (1966). *J. Am. Chem. Soc.* **88**, 3498.

Shams El Din, A. M. (1961). *Electrochim. Acta* **4**, 52.

Sluyters-Rehbach, M., and Sluyters, J. H. (1970). *Electroanal. Chem.* **4**, 3.

Smith, D. H. (1969). *Chem. Process.* **15** (9), 4.

Spencer, M. S. (1969). "Proceedings of Society of Electrochemistry Meeting", Chester, 20.

Sternberg, H. W., Markby, R. E., and Wender, I. (1963). *J. Electrochem. Soc.* **110**, 425.

Sternberg, H. W., Markby, R. E., Wender, I., and Mohilner, J. (1966). *J. Electrochem. Soc.* **113**, 1060.

Sternberg, H. W., Markby, R. E., Wender, I., and Monilner, D. H. (1969). *J. Am. Chem. Soc.* **91**, 4191.
Sundermeyer, W. (1965). *Angew. Chem. Int. Ed.* **4**, 222.
Tsutsumi, S., and Koyama, K. (1968). *Disc. Faraday Soc.* **45**, 247.
Van Dyne, R. P., and Reilley, C. N. (1972a). *Anal. Chem.* **44**, 142.
Van Dyne, R. P., and Reilley, C. N. (1972b). *Anal. Chem.* **44**, 158.
Van Dyne, R. P., and Reilley, C. N. (1972c). *Anal. Chem.* **44**, 153.
Vetter, K. J. (1967). "Electrochemical Kinetics Theoretical and Experimental Aspects", Academic Press.
Vincenz-Chodkowska, A., and Grabowski, Z. R. (1964). *Electrochim. Acta* **9**, 789.
von Stackelberg, M., and Stracke, W. (1949). *Z. Elektrochem.* **53**, 118.
Wawzonek, S. (1967). *Science* **155**, 39.
Wawzonek, S., and Wearring, D. (1959). *J. Am. Chem. Soc.* **81**, 2067.
Wawzonek, S., Berkey, R., Blaha, E. W., and Runner, M. E. (1955). *J. Electrochem. Soc.* **102**, 235.
Wawzonek, S., Duty, R. C., and Wagenknecht, J. H. (1964). *J. Electrochem. Soc.* **111**, 74.
Webb, J. L., Mann, C. K., and Walborsky, H. M. (1970). *J. Am. Chem. Soc.* **92**, 2042.
Weedon, B. C. L. (1960). *Adv. Org. Chem.* **1**, 1.
Weinberg, N. L., and Reddy, T. B. (1968). *J. Am. Chem. Soc.* **90**, 91.
Weinberg, N. L., and Weinberg, H. R. (1968). *Chem. Rev.* **68**, 449.
Weinberg, N. L., Hoffman, A. K., and Reddy, T. B. (1971). *Tetrahedron Lett.*, 2271.
Wisdom, N. E. (1969). "Proceedings of the Electrochemical Society Spring Meeting", New York, 338.
Woods, R. (1967). *Electrochim. Acta* **13**, 1967.
Woodward, R. B., and Hoffmann, R. (1969). *Angew. Chem. Int. Ed.* **8**, 781.
Wroblowa, H., Piersma, B., and Bockris, J. O'M. (1963). *J. Electroanal. Chem.* **6**, 401.
Závada, J., Krupička, J., and Sicher, J. (1963). *Coll. Czech. Chem. Comm.* 1664.
Zuman, P. (1967a). *J. Polarographic Soc.* **13**, 53.
Zuman, P. (1967b). "Substituent Effects in Polarography", Plenum Press.
Zuman, P. (1969). "The Elucidation of Organic Electrode Processes", Academic Press.
Zuman, P., and Tang, S. (1963). *Coll. Czech. Chem. Comm.* **28**, 829.
Zuman, P., Barnes, D., and Ryvolova-Kejharova, A. (1968). *Disc. Faraday Soc.* **45**, 205.

AUTHOR INDEX

Numbers in italics refer to the pages on which references are listed at the end of each article.

A

Abram, I. I., 96, *122*
Adams, R. N., 155, 161, 163, 164, 193, 211, 215, *220, 223, 224*
Adrian, F. J., 67, 121, *122, 123*
Akasaki, Y., *152*
Alford, J. R., 36, *52*
Allen, D. M., 146, 149, *151*
Allen, R. B., 89, *126*
Allison, A. C., 160, *224*
Anderson, J. D., 155, *220*
Ando, W., 119, *123*
Ankers, W. B., 119, *123*
Antheunis, D., 56, 84, *126*
Araki, M., 120, *126*
Arapakos, P. G., 175, *220*
Arnett, E. M., 34, 41, *51*
Aten, A. C., 212, *223*
Atkins, P. W., 121, *123*
Atlanti, P., 120, *123*
Avaca, A., 175, *220*
Avrutskaya, I. A., 189, *221*
Azumi, T., 61, *126*
Azzaro, M. E., 19, *26*

B

Baardman, F., 32, 40, 41, 43, *52*
Backhurst, J. R., 218, *220*
Badar, Y., 22, 23, *26*
Baggaley, A. J., 211, *220*
Bagotzky, V. S., 166, *220*
Baizer, M., 155, 188, 217, *220*
Baker, E. B., 45, *52*
Balasiewicz, M. S., 171, *222*
Baldin, V. I., 76, *127*
Baldwin, J. E., 116, 117, 118, 119, *123*
Bamford, C. H., 101, *123*
Banthorpe, D. V., 120, *123*
Barbas, J. T., 76, 112, *125*
Bargon, J., 56, 77, 82, 93, 94, 95, 110, *123, 125*

Barltrop, J. A., 140, *151*
Barnes, D., 179, *225*
Barnes, K. K., 155, *223*
Barry, C., 211, *220*
Bartell, L. S., 1, 5, 14, 15, 16, 17, 24, *26*
Bartlett, P. D., 43, *52*, 96, *123*
Barton, F. E., 82, *125*
Bastiansen, O., 21, *26*
Bastien, I. J., 45, *52*
Bauld, N. L., 211, *221*
Beck, F., 155, 217, 218, *220*
Becker, R. H., 118, *126*
Beletskaya, I. P., 115, *123*
Benecke, H. P., 118, *123*
Benkeser, R. A., 175, 215, *220*
Berger, S., 99, *123*
Berkey, R., 174, *225*
Bersohn, M., 43, *52*
Berson, J. A., 98, 114, *123*
Bertram, J., 158, 173, 184, *220*
Bertrand, R. D., 85, *128*
Bessard, J., 181, *220*
Bethell, D., 57, 75, 82, 100, 103, 114, *123*, 149, *152*
Bewick, A., 155, 175, *220*
Bhattacharyya, D. N., 176, *220*
Bickel, A. F., 30, 50, *52*, 137, *152*
Bicklehaupt, F., 114, *123*
Biellmann, J., 120, *123*
Bigeleisen, J., 19, *26*
Bilevich, K. A., 99, *123*
Bingham, R. C., 34, *52*
Blaha, E. W., 174, *225*
Blank, B., 57, 79, 85, 86, 87, 105, 106, *123*
Blomberg, C., 114, *123*
Blomgren, E., 186, *220*
Bobbitt, J. M., 192, *220*
Bockris, J. O'M., 166, 167, 169, 186, 194, 197, *220, 223, 225*
Bodewitz, H. W. H. J., 114, *123*
Boer, F. P., 51, *51*
Boháčková, V., 159, *223*
Bollinger, J. M., 184, *224*

Braithwaite, D., 173, 194, 218, *220*
Brand, P. A. T. M., 107, *124*
Brettle, R. 194, 211, *220*
Brière, R., 120, *123*
Brinkman, M. R., 57, 75, 82, 103, 114, *123*
Brokken-Zijp, J., 83, 85, 86, *126*
Brongersma, H. H., 50, *51*
Brookhart, M., 47, *51*, 131, *152*
Brouwer, D. M., 33, 35, 36, 37, 40, 41, 45, *51*
Brown, C., 119, *123*
Brown, E. R., 164, *220*
Brown, H. C., 18, 19, *26*, 43, 47, *51*
Brown, J. E., 116, 117, 118, *123*
Brown, O. R., 155, 161, 196, *220*
Bryce-Smith, D., 110, *123*, 137, *152*
Bubnov, N. N., 99, *123*
Buchachenko, A. L., 54, 76, 82, 83, 84, 85, 89, 90, 99, 115, *123, 124, 126, 127*
Buchanan, I. C., 121, *123*
Büchi, G., 142, *152*
Buck, H. M., 50, *51*
Burgess, E. M., 142, *152*
Buschow, K. H. J., 176, *220*
Bushby, R. J., 98, 114, *123*
Buteiko, Z. F., 115, *126*
Butler, R. M., 218, *220*
Bylina, G. S., 85, *124*
Byrd, L., 202, *221*

Clark, W. G., 101, *124*
Closs, G. L., 54, 56, 57, 75, 82, 95, 96, 98, 103, 105, 106, 108, 114, *124, 127*
Closs, L. E., 56, 57, 103, 108, *124*
Clusius, K., 3, *26*
Cocivera, M., 102, 106, 107, 108, 109, *124, 128*
Coetzee, J. F., 176, *221, 223*
Cole, T. M., Jr., 130, 132, 145, 146, 147, 148, 149, *152*
Coleman, J., 158, 184, *221*
Colpa, J. P., 121, *125*
Commeyras, A., 41, *52*
Conway, B. E., 169, 172, 194, 206, *221*
Cook, P. M., 119, *126*
Cooke, A. S., 22, 23, *26*
Cooper, R. A., 56, 57, 76, 77, 86, 88, 89, 111, 112, *124, 128*
Cordell, R. W., 117, 118, *123*
Corey, E. J., 211, *221*
Coulson, J. M., 218, *220*
Cox, R. H., 76, 112, 113, *125*
Cross, P., 42, *52*
Cross, P. C., 1, *27*
Cubbon, R. C. P., 82, *124*
Cuddy, B. D., 36, *52*
Cupus, C. A., 184, *224*
Czochralska, B., 171, *222*

C

Cabell, P. W., 131, 136, *152*
Cadogan, J. I. G., 98, *124*
Callister, J. D., 103, *123*
Calvin, M., 69, *123*
Campbell, P. H., 144, *152*
Carrington, A., 54, *124*
Carter, R. E., 4, 5, 21, 22, 23, 24, *26*
Casanova, J., 211, *221*
Case, B., *224*
Casson, J. E., 101, *123*
Cauquis, G., 155, 163, 181, 211, *220, 221*
Chang, J., 155, *221*
Chanon, M., 21, 25, *26*
Chapman, O. L., 142, *152*
Charlton, J. L., 77, *124*
Chemla, M., 194, *223*
Childs, R. F., 131, 133, 135, 137, 142, 144, *152*
Chiu, N. W. K., 144, *152*
Chloupek, F. J., 47, *51*
Christie, J. H., 164, *221*
Ciuhandu, G., 50, *52*
Clare, P. N., 151, *152*
Clark, D., 158, 165, 177, *221*

D

Dahlgren, L., 22, 23, 24, *26*
Dalton, J. C., 105, *128*
Damaskin, B. B., 159, *221*
Danen, W. C., 211, *221*
Dauben, H., Jr., 130, *152*
Davies, J. D., 176, 178, *224*
Dawes, K., 140, *151*
Day, A. C., 140, *151*
Decius, J. C., 1, *27*
Degner, D., 170, 188, *223*
De Kanter, F. J. J. J., 83, 85, 86, *126*
Dekkers, H. P. J. M., 50, *51*
Delahay, P., 161, 166, 167, 172, 183, *221*
DeMember, J. R., 41, *52*
den Hollander, J. A., 56, 67, 76, 77, 82, 83, 84, 107, *124, 125*
De Nova, V., 191, *221*
DeTar, D. F., 82, *124*
Deugau, K., 144, *152*
Deutch, J. M., 121, *124*
Devanathan, M. A. V., 167, *220*
Dieleman, J., 176, *220*
Dietz, R., 188, 212, *224*
Dodin, G., 175, *221*

Dodonov, V. A., 90, *127*
Dombchik, S. A., 83, *124*
DoMinh, T., 107, *124*
Doubleday, C. E., 108, *124*
Doughty, A., 174, *221*
Dubois, J. E., 175, *221*
Dumitreanu, A., 50, *52*
Dutkiewicz, E., 188, *221*
Duty, R. C., 196, *225*
Dvořak, J., 159, *223*

E

Eberson, L., 155, 172, 183, 189, 194, *221*
Eiben, E., 121, *124*
Eiszner, J. R., 102, *128*
Ellenbogen, P. E., 89, *126*
Elving, P. J., 162, 171, 194, *221*
Engel, P. S., 96, *123*
Entine, G., 69, *124*
Erickson, W. F., 117, 119, *123*
Ermanson, L. V., 99, *123*
Evans, G. T., 121, *126*
Evans, J. C., 45, *52*
Eyring, H., 16, *26*

F

Fahey, R. C., 15, *27*
Fahrenholtz, S. R., 84, 92, *124*
Faita, G., 201, *221*
Falbe, J., 29, *52*
Farenhorst, E., 137, *152*
Fawcett, W. R., 218, *220*
Feiner, S., 120, *128*
Feit, E. D., 105, *128*
Fenical, W., 149, *152*
Feokstistov, L. G., 170, 171, *221*
Ferruti, P., 69, *124*
Fessenden, R. W., 121, *124, 127*
Fichter, F., 155, *221*
Filipescu, N., 138, *152*
Fioshin, M. YA., 189, *221*
Fischer, H., 54, 56, 57, 73, 76, 79, 82, 84,
    85, 86, 87, 94, 95, 105, 106, 110, 121,
    123, 124, 125, 126
Fleischmann, M., 156, 158, 164, 165, 169,
    172, 174, 177, 178, 184, 189, 191, 194,
    196, 197, 201, 202, 211, 218, 129, *220,*
    221, 222
Flicker, M., 69, *125*
Foote, C. S., 141, *152*
Franklin, T. C., 166, *222*
Freed, J. H., 121, *127*

French, W. G., 192, *223*
Freudenberg, B., 98, *127*
Fritsch, J. M., 211, *224*
Frumkin, A. N., 168, 185, *222*
Fry, A. J., 163, *222*
Fry, J. L., 34, 36, *52*

G

Gaasbeek, C. J., 47, *52*, 131, 132, *152*
Galli, R., 169, 189, 194, *222*
Garst, J. F., 76, 82, 112, 113, *125, 127*
Geluk, H. W., 36, *52*
Genies, M., 163, *221*
Gerischer, H., 164, 218, *222*
Gibian, M. V., 84, *128*
Gilbert, A., 137, *152*
Gilchrist, T. L., 116, *125*
Gill, D., 69, *124*
Gillet, I. E., 173, 188, *222*
Gilman, S., 191, 192, 197, *222*
Giner, J., 197, *222*
Giumanini, A. G., 118, *126*
Glarum, S. H., 121, *125*
Glasstone, S., 189, *222*
Gleicher, G. J., 34, *52*
Goering, H. L., 43, *52*
Gol'dfarb, E. I., 83, 90, *126*
Goodridge, F., 217, 218, *220, 222*
Gordon, A. J., 4, 20, *26*
Gordon, L. G. M., 169, *221*
Gough, T. A., 160, *224*
Grabowski, Z. R., 171, *222, 225*
Graeve, R., 4, 20, *26*
Gragerov, I. P., 99, *126*
Grant, D. M., 85, *128*
Greeley, R., 130, 132, 145, 146, 147, *152*
Griffiths, J., 138, *152*
Grigat, E., 43, *52*
Grigor, A. F., 3, *26*
Groen, A., 106, 107, *128*
Grovenstein, E., 116, *125*
Guhn, G., 97, *125*
Gunning, H. E., 95, *128*
Gurd, R. C., 121, *123*
Gurney, R. W., 206, *222*
Guthke, H., 217, 218, *220*

H

Haber, F., 162, *222*
Hackler, R. E., 117, 119, *123*
Hagen, E. L., 40, 41, *52*
Hale, J., 211, *222*

Hale, J. M., 191, *223*
Halevi, E. A., 19, *26*
Halpern, Y., 184, *224*
Hamann, S. D., 204, *222*
Hampson, G. C., 21, *26*
Hampson, N. A., 191, 196, *222*
Hanack, M., 45, *52*
Hand, R., 200, *222*
Harris, J. M., 34, *52*
Harris, M. M., 22, 33, *26*
Harrison, J. A., 155, *220*
Hart, H., 138, 142, *152*
Hauff, S., 99, *123*
Hausser, K. H., 55, *125*
Hawley, M. D., 211, *221*
Hayes, J., 57, 75, 82, 103, 114, *123*
Headridge, J. B., 176, *222*
Hedrick, J. L., 176, *221*
Heesing, A., 97, *125*
Heitner, C., 24, *26*
Herring, C., 69, *125*
Hiatt, R., 82, *125*
Hickling, A., 189, *222*
Hilbers, C. W., 30, 50, *52*
Hilton, C. L., 162, *221*
Hine, K. E., 142, 144, *152*
Hirota, H., 69, *125*
Hoffman, A. K., 211, *225*
Hoffman, R. W., 97, 116, *125*
Hoffmann, R., *128*, 134, 135, 144, *153*, 211, *225*
Hogeveen, H., 30, 31, 32, 35, 36, 37, 39, 40, 41, 42, 43, 44, 45, 46, 47, 60, *51*, *52*, 131, 132, 135, 136, 142, *152*
Hoijtink, G. J., 176, 211, 212, *220*, *223*
Hollaender, J., 96, 97, *125*
Horányi, G., 191, 196, *223*
Horeau, A., 17, 18, *26*
Horiuti, J., 206, *223*
Horner, L., 170, 188, *223*
Houdard-Pereyre, J., 140, *152*
Howlett, K. E., 4, 5, 12, 21, 25, *26*
Hoyoshi, H., 69, *125*
Hudson, R. F., 119, *123*
Hui, M. M., 105, *128*
Humphrey, Jr., J. S., 16, *26*
Hurst, J. J., 142, *152*
Hush, N. S., 196, *223*, *224*
Hutchinson, D. A., 121, *125*

I

Ibl, N., 177, 215, *223*
Ignatenko, A. V., 89, 90, *125*
Il'yasov, A. V., 83, 90, *126*

Imai, I., 119, *123*
Ioffe, N. T., 99, *123*
Isaev, I. S., 134, *152*
Ismail, B. M., 218, *222*
Ito, H., 46, *52*
Itoh, K., 69, *125*
Iwamura, H., 94, 95, 119, 120, *125*
Iwamura, M., 54, 94, 95, 115, 119, 120, *125*

J

Jagi, H., 192, *220*
Jacobus, J., 115, *125*
Jagow, R. H., 15, *27*
Jansen, H. B., 50, *52*
Jemison, R. W., 118, *125*
Jesch, C., 186, *220*
Johnsen, V., 56, 82, *123*
Johnson, J. W., 197, *223*
Johnston, H. S., 213, *223*
Joslin, T., 201, *222*
Joussot-Dubien, J., 140, *152*
Junggren, E., 21, *26*
Jutz, C., 132, *152*

K

Kaiser, B. H., 97, *125*
Kaiser, E. M., 175, 215, *220*
Kaiser, E. T., 211, *221*
Kalinin, V. N., 95, 99, *126*
Kalinkin, M. I., 99, *123*
Kanakam, P., 171, *223*
Kantor, M. L., 142, *152*
Kaplan, E. D., 19, 20, *26*
Kaplan, F., 42, *52*
Kaplan, L., 139, *152*
Kaplan, M. L., 100, *125*
Karabatsos, G. J., 16, 17, *26*
Kaptein, R., 56, 57, 67, 68, 69, 73, 74, 75, 76, 77, 82, 83, 84, 85, 86, 87, 90, 91, 107, *125*
Kargin, J. H., 192, *223*
Kasukhin, L. F., 95, 115, *125*
Kato, M., 144, *152*
Kawazoe, Y., 120, *126*
Kearns, D. R., 149, *152*
Keizer, V. G., 36, *52*
Kellogg Co., 215, *223*
Kellogg, R. M., 142, *152*
Kensler, T. T., 211, *221*
Kessenikh, A. V., 54, 82, 83, 85, 89, 90, 115, *123*, *124*, *125*, *126*, *127*
Kinoshita, M., 61, *126*

Kiprianova, L. A., 99, *126*
Klein, H. S., 22, *27*
Klein, M. P., 69, 79, *124, 128*
Klopman, G., *44, 52*
Knapczyk, J. W., *152*
Kobrina, L. S., 84, *126*
Koch, D. F. A., 172, *223*
Koch, V. R., 184, *223*
Kochi, J. K., 121, *126*
Koehl, 194, 211, *223*
Koelling, J. G., *19, 26*
Koenig, T., 99, *126*
Kollmar, H., 184, *223*
Kolthoff, I. M., 176, *223*
Konovalov, A. G., 76, *128*
Koptyug, V. A., 134, *152*
Korchinskii, 171, *223*
Korinek, K., 172, 191, 196, 197, 202, *222*
Koryta, J., 159, *223*
Koyama, K., 177, *225*
Kramer, P., 44, *52*
Krebs, A. W., 145, *152*
Kreevoy, M. M., 16, *26*
Krupička, J., 171, *225*
Krusik, P. J., 121, *126*
Kuhn, S. J., 50, *52*
Kuindersma, K., 142, *152*
Kushida, K., 94, *125*
Kuwana, T., 192, *223*
Kuzubova, L. I., 134, *152*
Kwant, P. W., 135, *152*

L

La Londe, R. T., 211, *221*
Lam, L. K. M., 36, *52*
Lambert, R. F., 175, 215, *220*
Lamola, A. A., 110, *127*
Landau, R. L., 11, *126*
Landi, H. P., 197, *223*
Lane, A. G., 99, *126*
Langer, S. H., 197, *223*
Lansbury, P. T., 119, *126*
Lantelme, F., 194, *223*
Large, R. F., 155, 161, 164, *220, 221*
Larsen, J. W., 41, *51*
Larshcheid, M. E., 184, *223*
Lawler, R. G., 54, 57, 73, 76, 77, 86, 88, 89,
    110, 111, 112, 114, 115, 121, *124, 126,*
    127, 128
Lawless, J. G., 211, *221*
Lawson, A. J., 119, *123*
Le Duc, J. A. M., 215, *223*
Lee, C. C., 176, *220*

Lee, J. B., 191, 196, *222*
Leedy, D. W., 211, *224*
Leffek, K, T., 10, 14, 15, 16, 20, 24, *26*
Lehnig, M., 56, 73, 76, 84, 121, *125, 126*
Leibfritz, D., 99, *127*
Lemaire, H., 120, *123*
Lepley, A. R., 56, 85, 88, 110, 111, 118, 119,
    126, 128
Levich, V. G., 206, *223*
Levin, Y. A., 83, 90, *126*
Levit, A. F., 99, *126*
Lippmaa, E. T., 83, 85, 96, 99, *126*
Livingston, R., 121, *126, 128*
Llewellyn, J. A., 14, 15, 16, *26*
Lochmüller, C. H., 95, 96, 98, *127*
Loken, H. Y., 56, 76, 77, *128*
Longuet-Higgins, H. C., 137, *152*
Lown, J. W., 95, *128*
Lubinowski, J. J., *152*
Ludwig, U., 118, *128*
Lui, C. Y., 41, *52*
Lukas, J., 44, *52*, 184, *224*
Lund, H., 156, 162, *223*
Lustgarten, R. K., 47, *51*, 131, *152*

M

Mabey, W. R., 99, *126*
McBride, J. M., 98, 114, *123*
McDonald, G. J., 18, *26*
MacDonald, K. I., *222*
McEwen, W. E., *152*
McGlynn, S. P., 61, *126*
McGuire, D. K., 176, *221*
McIntyre, J. S., 45, *52*
McKervey, M. A., 36, *52*
Mackor, E. L., 30, 45, 50, *51, 52*
McLachlan, A. D., 54, *124*
McLauchlan, K. A., 121, *123*
MacLean, C., 30, 50, *52*
Mamatyuk, V. I., 83, *126*, 134, *152*
Mani, J., 149, *153*
Mann, C, K., 156, 173, 196, *223, 225*
Manning, G., 193, 215, *223*
Manoušek, O., 192, *223*
Mansfield, J. R., 169, *222*
Marcoux, L. S., 211, *224*
Marcus, M. F., 211, *221*
Marcus, R. A., 206, 213, *223*
Markby, R. E., 173, 175, *224, 225*
Marnett, L. J., 96, 98, *127*
Marshall, J. H., 121, *125*
Martin, A. J., 171, 194, *221*
Maruyama, K., 79, 109, 110, *126*
Maruyama, T., 79, 109, 110, *126, 127, 128*

Marzilli, T. A., 114, 115, *128*
Masui, M., 163, *223*
Matheson, A. F., 10, 20, *26*
Maurey, M., 211, *220, 221*
Mayeda, E. A., 173, 189, 190, *223*
Mayer, J. E., 4, 21, *27*
Mazzocchi, P. H., 143, *152*
Méchin, B., 114, *127*
Medvedev, B. Y., 99, *123*
Mehl, W., 191, *223*
Meinwald, J., 143, *152*
Meites, L., 161, 164, 179, *223*
Melander, L., 4, 5, 21, 23, *26*
Mels, S. T., 175, *220*
Mennitt, P. G., 79, 105, 106, *122*
Merz, E., 98, *127*
Metzger, J., 21, 25, *26*
Migita, T., 119, *123*
Mill, T., 96, *127*
Miller, I. J., 144, *152*
Miller, L. L., 162, 173, 184, 189, 190, 202, 211, *221, 223*
Milne, G. S., 96, *122*
Milovanovic, J., 82, *128*
Mislow, K., 4, 17, 18, 20, 21, *26*
Mitchell, T. N., 115, *127*
Mitnick, M., 163, *222*
Miura, I., 120, *125*
Mo, Y. K., 184, *224*
Modena, G., 45, *52*
Mohilner, J., 173, *224*
Monilner, D. H., 175, *225*
Morgan, K. J., 47, *51*
Morley, J. R., 191, 196, *222*
Morris, D. G., 117, 118, 119, *125, 127*
Morris, J. I., 76, 82, 112, 118, *125, 127*
Morrison, R. C., 76, 82, 112, *125, 127*
Mukai, T., *152*
Müller, K., 76, 82, 105, 106, *127*, 167, *220*
Murrell, J. N., 69, *127*
Mylonakis, S. G., 96, *128*

N

Nagakura, S., 69, *125*
Nagy, F., 191, 196, *223*
Nakaido, S., 119, *123*
Nakayama, J., 119, *125*
Nash, M. J., 142, *152*
Naulet, N., 114, *127*
Nelson, R. F., 220, 211, *222, 223, 224*
Neta, P., 121, *127*
Neumann, W. P., 96, 97, *125*
Nicholson, R. S., 164, *224*
Niederer, P., 99, *122, 127*

Nieh, M. T., 96, *128*
Nishida, T., 119, 120, *125*
Noguchi, I., 192, *220*
Nouaille, A., 17, 18, *26*
Noyes, R. M., 68, 72, *127*
Noyori, R., 144, *152*
Nürnberg, H. W., 186, *224*

O

O'Connor, J. J., 183, *224*
Oelderik, J. M., 37, 40, *51*
Ohashi, K., 46, *52*
Ohnishi, Y., 144, *152*
Okhlobystin, O. Y., 99, *123*
Olafson, B., 189, *221*
Olah, G. A., 41, 44, 45, 50, *52*, 175, 184, *224*
Oldfield, J., 172, 219, *222*
Oldham, K. B., 196, *223*
Olivani, F., 169, 189, 194, *222*
Oosterhoff, L. J., 50, *51*, 56, 57, 84, 87, *125, 126*
Opgenforth, H. J., 98, *127*
Ostermann, G., 117, 118, 119, *127, 128*
Otsuki, T., 109, 110, *126, 127, 128*
Owen, E. D., 146, 149, *151*

P

Pappiaonnou, C. G., 82, *128*
Papaioannou, C. G., 16, 17, *26*
Parker, V. D., 171, 183, 184, 193, 215, *221, 223, 224*
Parrington, B., 135, 137, *152*
Parsons, R., 170, 188, *221, 224*
Pathy, M. S. V., 171, *223*
Patsch, M., 118, *128*
Paul, H., 121, *125*
Paulson, D. R., 106, 108, *124*
Pavlik, J. W., 138, 139, *152*
Payne, R., 187, 188, *224*
Pearl, I. A., 183, *224*
Pedersen, J. B., 121, *127*
Pekhk, T. I., 83, 85, 99, *126*
Peover, M. E., 156, 158, 160, 176, 178, 181, 188, 211, 212, *224*
Percival, P. W., 121, *123*
Peters, J. A., 43, *52*
Petrii, O. A., 166, *224*
Petrov, I. N., 164, *222*
Petrovich, J. P., 155, *220*
Petrovskii, P. V., 99, *123*
Piekarski, S., 161, *224*

Piersma, B., 197, *225*
Pincock, R. E., 43, *52*
Pittman, C. U., 175, *224*
Pitts, J. N., Jr., 149, *153*
Pletcher, D., 155, 156, 158, 165, 169, 172, 174, 176, 177, 178, 184, 189, 191, 194, 196, 197, 210, 202, 211, *220, 221, 222, 224*
Plimley, R. E., 218, *220*
Pobedimskii, D. G., 83, 90, *126*
Polanyi, M., 206, *223*
Ponomarchuk, M. P., 95, 99, 115, *126*
Popp, G., 155, *221*
Porter, N. A., 95, 96, 98, *127*
Powell, J. S., 212, *224*
Prinstein, R., 42, *52*

R

Raber, D. J., 34, 36, *52*
Race, G. M., 178, 211, *222*
Radlick, P., 149, *152*
Rakshys, J. W., 112, 114, *127*
Rassat, A., 120, *123*
Rausch, D. J., 140, *153*
Reddy, T. R., 199, *225*
Reed, R. G., 163, *222*
Reeves, R. M., 186, *224*
Rei, M. H., 43, *51*
Reilley, C. N., 201, 202, 203, *225*
Remco, J. R., 121, *128*
Richey, H. G., 45, *52*
Richey, J. M., 45, *52*
Riddiford, A. C., 161, *224*
Riekena, E., 162, *223*
Rieker, A., 99, *123, 127*
Rifi, M. R., 162, 213, *224*
Roberts, R. D., 76, *125*
Robertson, R. E., 14, 15, 16, *26*
Rogers, T. R., 138, *152*
Roobeek, C. F., 32, 35, 36, 37, 39, 40, 41, 42, 43, 44, 45, 46, *52*
Ros, P., 50, *52*
Rosenfeld, J., 10, 41, *52*
Rosenthal, I., 171, 194, *221*
Rosewell, D. F., 149, *153*
Roth, H. D., 100, 101, 102, 110, *124, 125, 127*
Roussel, C., 21, 25, *26*
Rüchardt, C., 98, 99, *126, 127*
Rudd, E. J., 169, *221*
Runner, M. E., 174, *225*
Rykov, S. V., 54, 76, 82, 83, 84, 85, 89, 90, 94, 115, *123, 124, 125, 126, 127*
Ryvolova-Kejharova, A., 179, *225*

S

Sabol, M. A., 175, *220*
Saidashev, I. I., 83, 90, *126*
Sakai, M., 133, *152*
Saluvere, T., 99, *126*
Samitov, Y. Y., 83, 90, *126*
Sato, N., 194, *224*
Sato, S., 94, *125*
Saunders, M., 40, 41, *52*
Sayo, H., 163, *223*
Scanlon, B., 191, 196, *222*
Schäffer, H., 155, *221*
Scheppele, S. E., 16, 17, *26*
Scheraga, H. A., 16, 23, 24, *26*
Schewene, C. B., 43, *52*
Schlatmann, J. L. M. A., 36, *52*
Schleyer, P. v. R., 34, 36, *52*
Schlosberg, R. H., 44, *52*
Schöllkopf, U., 117, 118, 119, *128*
Schossig, J., 117, 119, *128*
Schuler, R. H., 121, *124, 127*
Schulman, E. M., 85, *128*
Schumacher, H., 130, 132, 145, 146, 147, *152*
Schwarz, W., 164, *224*
Scott, M. K., 175, *220*
Scott, R. A., 16, 23, 24, *26*
Scott, R. M., 117, 119, *123*
Sekine, T., 194, 195, *224*
Seltzer, S., 96, *128*
Selvig, A., 177, 215, *223*
Seo, T. E., 211, *224*
Serve, D., 181, *220*
Setser, D. W., 101, *124*
Shain, I., 164, *224*
Shams El Din, A. M., 197, *224*
Shen, J., 184, *224*
Shindo, H., 79, 109, 110, *126, 127, 128*
Shine, H. J., 120, *127*
Shiner, Jr., V. J., 16, *26*
Shiomi, K., 94, 95, 120, *125*
Shobataki, M., 96, 98, *127*
Sholle, V. D., 94, *127*
Shone, R. L., 16, 17, *26*
Shuster, D. I., 142, *152*
Sicher, J., 171, *225*
Siefert, E. E., 101, *124*
Simons, J. P., 104, *128*
Simpson, A. F., 121, *123*
Skaletz, D. H., 170, 188, *223*
Skell, P. S., 50, *52*
Slonim, I. Y., 76, *127*
Sluys-Vander Vlugt, M. J., 50, *51*
Sluyters, J. H., 164, *224*
Sluyters-Rehbach, M., 164, *224*

Smaller, B., 121, *128*
Smid, J., 176, *220*
Smith, D. H., 219, *224*
Smith, D. W., 82, *127*
Smith, H. O., 184, *223*
Solomon, B. S., 96, *122*
Sonnichsen, G. C., 16, 17, *26*
Sorensen, T. S., 144, *152*
Sothern, R. D., 166, *222*
Spencer, M. S., 192, *224*
Srinivasan, S. K., 166, 194, *220*
Starer, I., 50, *52*
Steel, C., 96, *122*
Steele, W. A., 3, *26*
Stegmann, H. B., 99, *127*
Stehlik, D., 55, *125*
Stern, M. J., 10, 19, *27*
Sternberg, H. W., 173, 175, *224, 225*
Stevens, G., 103, *123*
Stork, G., 43, *52*
Storr, R. C., 116, *125*
Stracke, W., 194, *225*
Strausz, O. P., 95, *128*
Streitwieser, Jr., A., 15, 22, *27*
Stringham, R. S., 96, *127*
Sugimoto, T., 140, *153*
Sugino, T., 194, 195, *224*
Sugita, N., 30, *52*
Summers, A. J. H., 140, *151*
Sundermeyer, W., 173, *225*
Sussman, D. H., 142, *152*
Suzuki, J., 119, *123*
Suzuki, S., 15, 27, 46, *52*
Szwarc, M., 176, *220*

T

Taguchi, V., 131, *152*
Takezaki, Y., 30, *52*
Takino, T., 142, *152*
Takuwa, A., 109, *127*
Tamura, M., 94, 120, *125*
Tan, C. C., 98, *127*
Tang, S., 179, 180, *225*
Taymaz, K., 120, *128*
Teixeira-Dias, J. J. C., 69, *127*
Tennakoon, C. T. K., 172, 219, *222*
Tezuka, T., *152*
Thirsk, H. R., 196, *220*
Thornton, B., 196, *220*
Thornton, E. R., 19, 20, *26*
Tickle, P., 103, *123*
Tolgyesi, W. S., 45, *52*
Tomkiewicz, M., 79, 106, 107, *128*
Tonellato, V., 45, *52*

Toyama, T., 119, *123*
Tozune, T., 119, *123*
Tremelling, M., 98, 114, *123*
Trifunac, A. D., 56, 57, 75, 95, 96, 103, 114, *124*
Trozzolo, A. M., 84, 92, 108, *124*
Tsolis, A., 96, *128*
Tsuda, Y., 163, *223*
Tsutsumi, S., 177, *225*
Turro, N. J., 105, *128*

U

Udupa, H. W. V., 171, *223*
Urman, Y. G., 76, *128*
Urry, W. H., 102, *128*

V

Van Bekkum, H., 43, *52*
Vance, C. J., 169, 172, 189, 194, *222*
Van Dyne, R. P., 201, 202, 203, *225*
van der Waals, J. H., 121, *128*
van Tamelen, E. E., 130, 132, 145, 146, 147, 149, *152*
Vassilyev, Yu., B., 166, *220*
Veeman, W. S., 121, *128*
Vegter, G. C., 31, *52*
Verheus, F. W., 87, *126*
Vertés, G., 191, 196, *223*
Vetter, K. J., 159, *225*
Vielstich, W., 164, *222*
Vijh, A. K., 169, 172, 194, *221*
Vincenz-Chodkowska, A., 171, *222, 225*
Vlasova, L. V., 83, *126*
Voithenleiter, F., 132, *152*
Vol'eva, V. B., 115, *123*
von Stackelberg, M., 194, *225*

W

Wagenknecht, J. H., 196, *225*
Wahl, Jr., G. H., 4, 20, *26*
Waits, H. P., 82, *128*
Walborsky, H. M., 196, *225*
Walling, C., 82, 84, 85, 88, *128*, 149, *153*
Wan, J. K. S., 121, *125*
Wang, H. H., 69, *124*
Ward, H. R., 54, 56, 57, 76, 77, 86, 88, 89, 110, 111, 112, 114, 115, *124, 127, 128*
Watanabe, H., 175, *220*
Wawzonek, S., 156, 174, 196 ,*225*
Wearring, D., 174, *225*

Webb, J. L., 196, *225*
Weedon, B. C. L., 184, *225*
Wei, K., 149, *153*
Weigand, K., 3, *26*
Weinberg, H. R., 156, *225*
Weinberg, N. L., 156, 199, *225*
Weisgraber, K. H., 192, *220*
Weissberger, A., 21, *26*
Weissman, S. I., 69, *125*
Wender, I., 173, *175*, *224*, *225*
Wentworth, G., 116, *125*
Werner, R., 98, 99, *126*, *127*
Westheimer, F. H., 4, 21, *27*
White, A. M., 41, *52*
White, E. H., 149, *153*
Whitham, G. H., 142, *152*
Whittaker, D., 103, *123*
Wieko, J., 149, *153*
Wieland, H., 149, *153*
Wigfield, D. C., 120, *128*
Wikel, J. H., 118, *123*
Willard, G. F., 119, *126*
Williams, G. H., 91, *128*
Wilson, E. B., Jr., 1, *27*
Wilt, J. W., 102, *128*
Wilzbach, K. E., 139, 140, *152*, *153*
Winstein, S., 47, *51*, 131, 133, *152*
Winter, J. G., 120, *123*
Wisdom, N. E., 184, *225*
Wolf, J. F., 184, *223*

Wolfsberg, M., 10, 19, *26*, *27*
Wong, S. K., 121, *125*
Woods, R., 172, *223*, *225*
Woodward, R. B., 116, *128*, 134, 135, 144, *153*, 211, *225*
Wroblowa, H., 197, *223*, *225*
Wynne-Jones, W. F. K. Lord, 164, 169, *222*

Y

Yagihara, T., 119, *123*
Yamura, A., 195, *224*
Yang, N. C., 105, *128*
Yankelevich, A. Z., 83, *126*
Yasutomi, T., 30, *52*
Yenida, S., 140, *153*
Yoshida, L., 140, *153*
Yoshida, M., 119, *125*
Yurchak, S., 197, *223*

Z

Závada, J., 171, *225*
Zeldes, H., 121, *126*, *128*
Zhidomirov, G. M., 54, *124*
Zuman, P., 178, 179, 180, 192, 209, 210, *223*, *225*

CUMULATIVE INDEX TO AUTHORS

Anbar, M., **7**, 115
Bell, R. P., **4**, 1
Bennett, J. E., **8**, 1
Bentley, T. W., **8**, 151
Bethell, D., **7**, 153; **10**, 53
Brand, J. C. D., **1**, 365
Brinkman, M. R., **10**, 53
Brown, H. C., **1**, 35
Cabell-Whiting, P. W., **10**, 129
Cacace, F., **8**, 79
Carter, R. E., **10**, 1
Collins, C. J., **2**, 1
Crampton, M. R., **7**, 211
Fendler, E. J., **8**, 271
Fendler, J. H., **8**, 271
Ferguson, G., **1**, 203
Fields, E. K., **6**, 1
Fleischmann, M., **10**, 155
Frey, H. M., **4**, 147
Gilbert, B. C., **5**, 53
Gillespie, R. J., **9**, 1
Gold, V., **7**, 259
Greenwood, H. H., **4**, 73
Hogeveen, H., **10**, 129
Johnson, S. L., **5**, 237
Johnstone, R. A. W., **8**, 151
Kohnstam, G., **5**, 121
Kreevoy. M. M., **6**, 63
Long, F. A., **1**, 1
Maccoll, A., **3**, 91
McWeeny, R., **4**, 73
Melander, L., **10**, 1
Mile, B., **8**, 1
Miller, S. I., **6**, 185
Modena, G., **9**, 185
More O'Ferrall, R. A., **5**, 331
Norman, R. O. C., **5**, 53
Olah, G. A., **4**, 305
Parker, A. J., **5**, 173
Peel, T. E., **9**, 1
Perkampus, H. H., **4**, 195
Pittmann, C. U., Jr., **4**, 305
Pletcher, D., **10**, 155
Ramirez, F., **9**, 25
Rappoport, Z., **7**, 1
Reeves, L. W., **3**, 187
Robertson, J. M., **1**, 203

Samuel, D., **3**, 123
Schaleger, L. L., **1**, 1
Scheraga, H. A., **6**, 103
Shatenshtein, A. I., **1**, 156
Silver, B. L., **3**, 123
Simonyi, M., **9**, 127
Stock, L. M., **1**, 35
Symons, M. C. R., **1**, 284
Thomas, A., **8**, 1
Tonellato, U., **9**, 185
Tüdös, F., **9**, 127
Turner, D. W., **4**, 31
Ugi, I., **9**, 25
Ward, B., **8**, 1
Whalley, E., **2**, 93
Williams, J. M., Jr., **6**, 63
Williamson, D. G., **1**, 365
Wolf, A. P., **2**, 201
Zollinger, H., **2**, 163
Zuman, P., **5**, 1

CUMULATIVE INDEX OF TITLES

Abstraction, hydrogen atom, from O–H bonds, **9**, 127
Acid solutions, strong, spectroscopic observation of alkylcarbonium ions in, **4**, 305
Acids, reactions of aliphatic diazo compounds with, **5**, 331
Activation, entropies of, and mechanisms of reactions in solution, **1**, 1
Activation, heat capacities of, and their uses in mechanistic studies, **5**, 121
Activation, volumes of, use for determining reaction mechanisms, **2**, 93
Aliphatic diazo compounds, reactions with acids, **5**, 331
Alkylcarbonium ions, spectroscopic observation in strong acid solutions, **4**, 305
Ammonia, liquid, isotope exchange reactions of organic compounds in, **1**, 156
Aromatic substitution, a quantitative treatment of directive effects in, **1**, 35
Aromatic substitution reactions, hydrogen isotope effects in, **2**, 163
Aromatic systems, planar and non-planar, **1**, 203
Arynes, mechanisms of formation and reactions at high temperatures, **6**, 1
A-S_E2 reactions, developments in the study of, **6**, 63

Base catalysis, general, of ester hydrolysis and related reactions, **5**, 237
Basicity of unsaturated compounds, **4**, 195
Bimolecular substitution reactions in protic and dipolar aprotic solvents, **5**, 173

Carbene chemistry, structure and mechanism in, **7**, 163
Carbon atoms, energetic, reactions with organic compounds, **3**, 201
Carbon monoxide, reactivity of carbonium ions towards, **10**, 29
Carbonium ions (alkyl), spectroscopic observation in strong acid solutions, **4**, 305
Carbonium ions, gaseous, from the decay of tritiated molecules, **8**, 79
Carbonium ions, photochemistry of, **10**, 129
Carbonium ions, reactivity towards carbon monoxide, **10**, 29
Carbonyl compounds, reversible hydration of, **4**, 1
Catalysis, general base and nucleophilic, of ester hydrolysis and related reactions, **5**, 237
Catalysis, micellar, in organic reactions: kinetic and mechanistic implications, **8**, 271
Cations, vinyl, **9**, 185
Chemically induced dynamic nuclear spin polarization and its applications, **10**, 53
CIDNP and its applications, **10**, 53
Conformations of polypeptides, calculations of, **6**, 103
Conjugated molecules, reactivity indices in, **4**, 73

Diazo compounds, aliphatic reactions with acids, **5**, 331
Dipolar aprotic and protic solvents, rates of bimolecular substitution reactions in, **5**, 173
Directive effects in aromatic substitution, a quantitative treatment of, **1**, 35

Electrode processes, physical parameters for the control of, **10**, 155
Electron spin resonance, identification of organic free radicals by, **1**, 284
Electron spin resonance studies of short-lived organic radicals, **5**, 53
Electronically excited molecules, structure of, **1**, 365
Energetic tritium and carbon atoms, reactions of, with organic compounds, **2**, 201
Entropies of activation and mechanisms of reactions in solution, **1**, 1
Equilibrium constants, N.M.R. measurements of, as a function of temperatures, **3**, 187
Ester hydrolysis, general base and nucleophilic catalysis, **5**, 237
Exchange reactions, hydrogen isotope, of organic compounds in liquid ammonia, **1**, 156

Exchange reactions, oxygen isotope, of organic compounds, 3, 123
Excited molecules, structure of electronically, 1, 365

Free radicals, identification by electron spin resonance, 1, 284
Free radicals and their reactions at low temperature using a rotating cryostat, study of, 8, 1

Gaseous carbonium ions from the decay of tritiated molecules, 8, 79
Gas-phase heterolysis, 3, 91
Gas-phase pyrolysis of small-ring hydrocarbons, 4, 147
General base and nucleophilic catalysis of ester hydrolysis and related reactions, 5, 237

H_2O–D_2O mixtures, protolytic processes in, 7, 259
Heat capacities of activation and their uses in mechanistic studies, 5, 121
Heterolysis, gas-phase, 3, 91
Hydrated electrons, reactions of, with organic compounds, 7, 115
Hydration, reversible, of carbonyl compounds, 4, 1
Hydrocarbons, small-ring, gas-phase pyrolysis of, 4, 147
Hydrogen atom abstraction from O–H bonds, 9, 127
Hydrogen isotope effects in aromatic substitution reactions, 2, 163
Hydrogen isotope exchange reactions of organic compounds in liquid ammonia, 1, 156
Hydrolysis, ester, and related reactions, general base and nucleophilic catalysis of, 5, 237

Ionization potentials, 4, 31
Isomerization, permutational, of pentavalent phosphorus compounds, 9, 25
Isotope effects, steric, experiments on the nature of, 10, 1
Isotope effects, hydrogen, in aromatic substitution reactions, 2, 163
Isotope exchange reactions, hydrogen, of organic compounds in liquid ammonia, 1, 150
Isotope exchange reactions, oxygen, of organic compounds, 3, 123
Isotopes and organic reaction mechanisms, 2, 1

Kinetics, reaction, polarography and, 5, 1

Mass spectrometry, mechanism and structure in: a comparison with other chemical processes, 8, 152
Mechanism and structure in carbene chemistry, 7, 153
Mechanism and structure in mass spectrometry: A comparison with other chemical processes, 8, 152
Mechanisms, organic reaction, isotopes and, 2, 1
Mechanisms, reaction, use of volumes of activation for determining, 2, 93
Mechanisms of formation and reaction of arynes at high temperatures, 6, 1
Mechanisms of reactions in solution, entropies of activation and, 1, 1
Mechanistic studies, heat capacities of activation and their uses in, 5, 121
Meisenheimer complexes, 7, 211
Micellar catalysis in organic reactions: kinetic and mechanistic implications, 8, 271

N.M.R. measurements of reaction velocities and equilibrium constants as a function of temperature, 3, 187
Non-planar and planar aromatic systems, 1, 203
Nuclear magnetic resonance, see N.M.R.
Nucleophilic catalysis of hydrolysis and related reactions, 4, 237
Nucleophilic vinylic substitution, 7, 1

O–H bonds, hydrogen atom abstraction from, 9, 127
Oxygen isotope exchange reactions of organic compounds, 3, 123

Permutational isomerization of pentavalent phosphorus compounds, 9, 25
Phosphorus compounds, pentavalent, turnstile rearrangement and pseudorotation in permutational isomerization, 9, 25
Photochemistry of carbonium ions, 10, 129
Planar and non-planar aromatic systems, 1, 203
Polarizability, molecular refractivity and, 3, 1
Polarography and reaction kinetics, 5, 1
Polypeptides, calculations of conformations of, 6, 103
Protic and dipolar aprotic solvents, rates of bimolecular substitution reactions in, 5, 173
Protolytic processes in H_2O–D_2O mixtures, 7, 259
Pseudorotation in isomerization of pentavalent phosphorus compounds, 9, 25
Pyrolysis, gas-phase, of small-ring hydrocarbons, 4, 147

Radicals, organic free, identification by electron spin resonance, 1, 284
Radicals, short-lived organic, electron spin resonance studies of, 5, 53
Reaction kinetics, polarography and, 5, 1
Reaction mechanisms, use of volumes of activation for determining, 2, 93
Reaction mechanisms in solution, entropies of activation and, 1, 1
Reaction velocities and equilibrium constants, N.M.R. measurements of, as a function of temperature, 3, 187
Reactions of hydrated electrons with organic compounds, 7, 115
Reactivity indices in conjugated molecules, 4, 73
Refractivity, molecular, and polarizability, 3, 1
Resonance, electron-spin, identification of organic free radicals by, 1, 284
Resonance, electron-spin, studies of short-lived organic radicals, 5, 63

Short-lived organic radicals, electron-spin resonance studies of, 5, 53
Small-ring hydrocarbons, gas-phase pyrolysis of, 4, 147
Solution, reactions in, entropies of activation and mechanisms, 1, 1
Solvents, protic and dipolar aprotic, rates of bimolecular substitution reactions in, 5, 173
Spectroscopic observation of alkylcarbonium ions in strong acid solutions, 4, 305
Steric isotope effects, experiments on the nature of, 10, 1
Stereoselection in elementary steps of organic reactions, 6, 185
Structure and mechanism in carbene chemistry, 7, 153
Structure of electronically excited molecules, 1, 365
Study of free radicals and their reactions at low temperatures using a rotating cryostat, 8, 1
Substitution, aromatic, a quantitative treatment of directive effects in, 1, 35
Substitution reactions, bimolecular, in protic and dipolar aprotic solvents, 5, 173
Substitution reactions, aromatic, hydrogen isotope effects in, 2, 163
Superacid systems, 9, 1

Temperature, N.M.R. measurements of reaction velocities and equilibrium constants as a function of, 3, 187
Tritiated molecules, gaseous carbonium ions from the decay of, 8, 79
Tritium atoms, energetic, reactions with organic compounds, 2, 201
Turnstile rearrangement in isomerization of pentavalent phosphorus compounds, 9, 25

Unsaturated compounds, basicity of, 4, 195

Vinyl cations, 9, 185
Volumes of activation, use of, for determining reaction mechanisms, 2, 93